Understanding the Financial Score

Understanding the Financial Score
Henry E. Riggs
www.morganclaypool.com

ISBN-10: 1598291688   paperback
ISBN-13: 9781598291681   paperback

ISBN-10: 1598291696   ebook
ISBN-13: 9781598291698   ebook

DOI 10.2200/S00074ED1V01Y200612TME001

A Publication in the Morgan & Claypool Publishers Series
SYNTHESIS LECTURES ON TECHNOLOGY, MANAGEMENT AND ENTREPRENEURSHIP #1

Lecture #1
Series Editor: Henry E. Riggs, President Emeritus of the Keck Graduate Institute of Applied Life Sciences

First Edition
10 9 8 7 6 5 4 3 2 1

Printed in the United States of America

Understanding the Financial Score

Henry E. Riggs
President Emeritus of the Keck Graduate
Institute of Applied Life Sciences

*SYNTHESIS LECTURES ON TECHNOLOGY, MANAGEMENT AND
ENTREPRENEURSHIP #1*

MORGAN & CLAYPOOL PUBLISHERS

ABSTRACT

Financial statements and information drawn from them confront us daily: in the media, in corporate annual reports, in the treasurer's reports for clubs or religious groups, in documents provided to employees and managers, as one considers alternative investments, in documents provided by homeowners' association and government agencies

Various readers of a company's "financial score" make decisions based on financial information: the company's managers devise actions to improve operations; investors buy or sell the corporation's securities; creditors decide how much to lend; customers judge the reliability of this supplier; potential employees decide whether to invest their careers in the company.

If you are training to be an accountant, find another book. This book's objective is to increase your ability to draw useful information from financial statements, and thus to make better decisions—in both your personal life and your professional life. Studying this book should help you be a better manager. That is both its objective and its perspective.

The book starts at square one; it assumes no prior knowledge on your part. To increase your financial literacy, you will learn the common nomenclature (but not esoteric jargon) used by accountants and financial experts. You will be equipped to ask insightful questions of experts, to engage them and your colleagues in thoughtful debates about financial and accounting issues, and to make better decisions.

KEYWORDS

Management, entrepreneurship, financial statement analysis, accounting, budgeting, regulations, debt leverage, financing corporations, financial decision-making, valuation, cash flow, sustainable growth, not-for-profit accounting

Contents

Preface

You know, or you should soon realize, that for a host of reasons you need to understand the financial score. Presumably, that is why you are now reading this preface. Financial statements themselves, and information drawn from them, confront you daily: in the media, in corporate annual reports, in the treasurer's report for your club or religious group, in documents you receive as an employee and manager, as you consider alternative ways you might invest your savings, in documents you receive from your homeowners' association, in government reports.

A primary reason for keeping the score at an athletic event is to determine which team won. Our reasons for keeping the financial score are more complex. Various readers of the "financial score" make decisions based on financial information: the company's managers devise actions to improve operations; investors buy or sell the corporation's securities; creditors decide how much to lend; customers judge the reliability of this supplier; potential employees decide whether to invest their careers in the company. These decisions require more detailed information than simply the bottom line, who won. As the baseball team manager uses data on batting averages, RBIs, errors, running speed, and home runs to decide the starting nine players and their batting order, the many readers of financial statements need lots of data and information to evaluate both the financial position and the financial performance of the corporation.

If you are training to be an accountant, find another book. This book has a different objective: to increase your ability to draw useful information from financial statements, and thus to make better decisions—in both your personal life and your professional life. You are a manager regardless of your occupation, indeed even if you are not employed. You manage your own financial resources; you assess the financial score of groups or organizations to which you belong; you manage your votes; you manage your professional and/or volunteer career.

Studying this book should help you be a better manager. That is both its objective and its perspective. It assumes that you are primarily interested in understanding financial statements of for-profit corporations, but the Appendix focuses on not-for-profit institutions and ways in which their financial statements differ from their for-profit cousins' statements.

Bear in mind that corporate financial statements are prepared for, and provided to, a wide variety of audiences. The paramount audience is the corporation's own managers—from the top (the chief executive and board of directors) to the bottom (the first-line supervisors) of the hierarchy. Financial statements are provided to managers not simply to quench their curiosities; these reports are designed to provide information that helps them make better decisions.

These decisions will affect the future of the corporation; they cannot affect the past. Yet financial statements are historical documents; they display what has happened, not what will happen. But if managers do not have a thorough understanding of the road that the corporation has traveled, they are unlikely to make sound decisions about the road ahead.

But what about other audiences? Shareholders and potential shareholders, as well as the army of brokers, security analysts and other advisors who counsel investors, are avid readers of financial statements. So are the customers and potential customers: how financially viable is this supplier of goods or services? So are the corporation's creditors such as banks: how confident can they be that the corporation will pay interest and repay principal on schedule? So are the employees and their representatives (labor unions): how secure is employment and can the corporation afford higher wages? And, of course, do not forget the tax collectors!

Think about all these questions for a moment. Can we expect financial statements to provide complete answers to all these complex questions? No, not complete answers, but certainly information essential to arriving at those answers. Financial statements can only provide information in monetary terms, and yet much that is important about the past and future of the company cannot be expressed in monetary terms. A key invention or a very successful sales call or the hiring of a brilliant marketing executive may be wildly significant for the company, yet the positive financial impact of these events may not be realized for months or even years. A key customer is lost or a lawsuit is filed against the company or a competitor lures away a key scientist; these too are important events, but their negative financial impacts may not be apparent for some time. A security analyst or potential customer or banker may commence his or her investigation of a particular corporation by reviewing its financial statements but the wise investigator will typically want to gather lots of other information, much of it nonmonetary.

This book starts at square one; it assumes no prior knowledge on your part, although I recognize that few of you are total neophytes in financial matters. To increase your financial literacy, you will learn the common nomenclature used by accountants and financial experts (but not the esoteric jargon that pervades all specialized professions); this learning is reinforced by bold highlighting of key terms when they first appear in the text, and a listing of those key terms at the end of each chapter. You will be equipped to ask insightful questions of experts and, indeed, to engage them and your colleagues in thoughtful debates about financial and accounting issues.

We begin in Chapters 1 and 2 with a review of the *structure* of the key financial statements—the balance sheet and the income statement—and the linkage between them. This structure is used just about universally (with slight variations in some other countries); it is essentially noncontroversial. The controversies, arguments, and temptations for accounting misbehavior arise in *valuing* transactions and conditions at the corporation and then deciding the *timing* of when those values should be reflected in the financial statements—as we shall explore in Chapters 3 and 4.

Chapter 5 puts a toe in the complicated waters of how corporations are financed, because financing decisions shape financial results, and thus statements, in major ways. Incidentally, you may be inclined to think that corporations, like consumers, should avoid borrowing or, alternatively, should borrow all they can. For the moment, please reserve judgment on that matter.

In Chapter 6 we explore a widely used financial statement, one derived from the balance sheet and income statement and focused on the flow of cash in and out of the corporation. Unsurprisingly, it is referred to as the cash flow statement! Why a separate statement? Because, as the saying goes and we shall see, cash is king!

In Chapter 7 we get to the specific and rather straightforward techniques used to analyze the three financial statements. Because the *relationships* between and among various values appearing in these statements are keys to understanding, we will focus on calculating and interpreting financial *ratios*.

For some of you, Chapter 7 may be as far as you want to go. Chapter 8 puts another toe in the water, this time of cost accounting, because the *costing* of products, projects, and services in so pervasive in our modern world: think about auto repair shops, professional service firms such as lawyers and architects, to say nothing of the task of determining what it costs to manufacture a car, a pencil, or a gallon of gasoline.

Chapter 9 looks to the future. Like financial statements, budgets and forecasts are pervasive in our lives. A solid understanding of an entity's financial history (i.e., its financial statements) is essential to reliable and useful budgeting and forecasting. This chapter considers some techniques for budgeting revenues and expenses, forecasting future financial statements, and guarding against near-term cash insolvency.

Finally, in our contemporary world, one is almost compelled to conclude any study of the financial score by highlighting some financial reporting misbehavior that has come to light in corporate America, particularly over the past decade. I hope that this final chapter will leave you with a healthy skepticism—but not cynicism—about what can be learned from and what might reasonably be questioned in financial statements that you review. Certainly, it should reinforce in your mind that the integrity of an organization's management is the key to assuring the soundness and reliability of its financial statements.

Because not-for-profit enterprises—education, social services, religious, and more—are a part of nearly all of our contemporary lives, the Appendix points out some of the challenges in understanding the financial score of not-for-profits compared to for-profits.

Henry E. Riggs
Stanford, California

CHAPTER 1

The Balance Sheet
Position not Performance

The two fundamental documents that present the financial score are the **balance sheet** and the **income statement**—the first discussed in this chapter and the second in Chapter 2. They are the primary products of the accounting system. In a sense the balance sheet is the more "fundamental" of the two because, as we shall see, it would be possible, though quite undesirable, to operate an accounting system with solely a balance sheet.

THE FUNDAMENTAL ACCOUNTING EQUATION

The form of the balance sheet is **Assets = Liabilities + Owners' Equity**. Think of assets, the left-hand side of the equation, as including everything the corporation owns, both physical things such as machinery, inventory and cash, and intangible things such as patents and trademarks. The right-hand side is everything the corporation owes. Thus, the equation can be translated into Owns = Owes.

The Owes side needs a little more explanation. Surely the company owes its liabilities, whether the amounts are owed to suppliers (goods and services bought on credit), to banks (loans), to taxing agencies for taxes not yet paid, or to employees for wages and salaries earned but not yet paid to them. But Owners' Equity is another matter.

Consider a simple algebraic rearranging of the balance sheet equation: Assets-Liabilities = Owners' Equity. Now it becomes clear why Owners' Equity is very often referred to as **Net Worth**—the difference between assets and liabilities is one measure of the "worth" of the corporation to its shareholders. Indeed, for the corporate form of business (as distinct from partnerships and sole proprietorships), owners' equity is properly referred to as **Shareholders' Equity**.

But you and I know that corporations do not "owe" their shareholders in the same sense that they owe their creditors. Still, corporations have an obligation to their shareholders: not to return their money (as with creditors) but to utilize the shareholders' funds effectively so that the shareholders ultimately benefit from receiving dividends and/or from an increase in the market value of their shares of stock.

Another way to think about this equation is as follows: Investments = Sources of Funds. The assets represent the investments of the funds that were provided (entrusted) to the corporation by its creditors and its shareholders. This form of the equation emphasizes the equality: the corporation cannot own more or less than the funds available to it to invest. The corporation's assets (including cash, remember) must be equal to the sum of its liabilities and shareholders' equity.

A STATEMENT OF FINANCIAL POSITION

Think of the balance sheet as a snapshot, as of a particular date, of what the corporation owns and owes (and those are necessarily equal). A balance sheet carries a specific date and shows the financial position of the corporation as of that date: the value of what the company owns, what it owes to its various creditors, and therefore what its shareholders' equity (or worth) is on that date. Its financial position will be different on the following date—perhaps not much different, but different nonetheless. Thus, the balance sheet is a Statement of Financial Position.

The balance sheet can tell you nothing about the financial performance of the company. You cannot tell by looking at the balance sheet whether the corporation's revenues (that is, sales) are large or small, growing or shrinking, nor whether the company is currently profitable or incurring losses. For information on those matters, you need to hold on until Chapter 2 where we discuss the income statement.

But do not denigrate the usefulness of the balance sheet. It tells us more than simply the total value of the assets; it tells us the value of various types of assets. You are going to feel somewhat better about a corporation that has lots of cash than one that has almost no cash. You will want to evaluate the amounts owed by customers—**Accounts Receivable (A/R)**. You recognize that these will probably be collected (turned into cash) relatively soon; on the other hand, if the accounts receivable balance is very high in relationship to the corporation's rate of sales, this high balance may suggest that the corporation is having trouble collecting from its customers. You can go through an analogous line of reasoning in evaluating the corporation's investment in **Inventories**.

And on the other side of the equation, an evaluation of the composition of the liabilities can be revealing. How much is owed to vendors/suppliers (**Accounts Payable, A/P**)—an appropriate amount, given the volume of business the corporation is doing, or is there evidence that the corporation is having trouble meeting the payment terms of its suppliers? Liabilities represent a demand on cash, and thus the timing of when these liabilities must be discharged, and how, can be all important.

Exhibit 1.1: *Typical Balance Sheet*

ASSETS	LIABILITIES & OWNERS' EQUITY
Current Assets	Current Liabilities
. Cash	Accounts payable
Accounts receivable	Wages & salaries payable
Inventory	Accrued liabilities
Prepaid expenses	Income taxes payable
Other current assets	Other current liabilities
Total current assets	Total current liabilities
Fixed Assets (Net)	Long-Term Debt
Factory equipment	Other Long-Term Liabilities
Office equipment	
Land & buildings	Shareholders' Equity
Total fixed assets	Invested capital
Intangible Assets	Retained earnings
Goodwill	Total shareholders' equity
Total Assets	Total Liabilities and Owners' Equity

STRUCTURE OF THE BALANCE SHEET

Exhibit 1-1 lists the categories of assets, liabilities, and owners' equity that might appear on a typical balance sheet of a merchandising or manufacturing company. Some of these categories could be combined, or, alternatively, further divided, as the corporation deems that more or less detail would be useful to various audiences for its financial statements. For example, the corporation may decide it is unnecessary to separate its **Fixed Assets** into factory equipment, office equipment, and land and buildings; these could all be combined into a single category. Or, alternatively, it might further categorize its cash into cash in the bank and cash-equivalent securities.

Assets

These categories are not listed in random order, but rather in order of decreasing **liquidity**. The more liquid the asset, the closer it is to becoming cash.

Of course, no asset is more liquid than cash itself, so cash is the first asset listed. Accounts receivable appear second because presumably the company expects to collect these amounts from customers within the next few months, although of course a small percentage of these accounts receivable may prove to be simply uncollectible. Inventory is next, as it has to be sold—that is, turned into accounts receivable—before it becomes cash.

Please do not worry about Prepaid Expenses at this point; we will attack that category in Chapter 4. In most every list we encounter, financial or otherwise, there seems always to be an "other" category; such is the case three times in Exhibit 1-1.

Now we encounter a subtotal: **Current Assets**, the sum of the first five categories. The temporal definition of "current" is 1 year, and included in current assets are those assets that either are now cash or will be turned into cash within the next 12 months. For example, if such extended payment terms have been provided to some customers that a portion of the accounts receivable are not due within the next year, that portion should be listed as a long-term asset (i.e., below the current asset subtotal).

The meaning of fixed assets is well explained by a phrase that is often used in its stead: Property, Plant, and Equipment. These assets, used in the operation of the business, typically have reasonably long useful lives, but not infinite lives. Thus, we will see in Chapter 4 that we have a method of reducing the value of these assets across their useful lives; you undoubtedly have heard the relevant term: depreciation. Note in Exhibit 1-1 the word "Net" that follows the label, Fixed Assets. That word signals to the reader that the values assigned to the fixed assets have been adjusted to reflect their declining value with age. Just how depreciation is calculated we leave until later.

You might be inclined to argue that certain of the fixed assets are readily saleable (for example, forklift trucks or salespersons' automobiles), could surely be turned into cash within the next 12 months, and therefore should be considered current assets. Yes, they "could" be sold, but that presumably is not the corporation's intent; the corporation bought these assets to use, not to resell. They are properly classified as long-term assets. On the other hand, suppose the company has made the decision to sell its fleet of salespersons' vehicles (and insist that the salespersons purchase their own vehicles and be reimbursed for mileage) and is now actively soliciting bids from prospective purchasers, with the expectation that the fleet sale will be consummated within the next few months. Then, indeed, the fleet should appropriately be considered a current asset.

Intangibles are "rights," not physical things. For example, a corporation might acquire a patent or a trademark, which it would then value as an intangible. Intangibles are often the key to the corporation's future profitability. Valuing these intangibles is perplexing, particularly when, as in the case of trademarks, the value has built up over many years of use. Think, for example, of the trademarks Pepsi, Google, HP, Apple, Nabisco, Pampers. Several of these trademarks could be sold for very large amounts, and yet are valued by their current owners at or near zero. We will see in Chapter 3 that intangibles are just one category on a very long list of assets and liabilities that we are severely challenged to value appropriately.

Goodwill is rather an oddball asset; it surely is not a tangible asset and, unlike some intangible assets, it cannot be sold or traded. It arises when one company buys another company, say, Federated Department Stores buys Macy's or Marshall Field's for a price that exceeds the value of the assets it obtains from the acquired company. If Federated pays out cash (reduces its asset Cash) by more than the values by which it increases its other assets (such as inventory and property, plant, and equipment), how do we make the balance sheet balance? By declaring the difference as Goodwill. Why would Federated pay more than the value of the assets it acquires? Because it believes that the future prospects for the operations being acquired—call it goodwill—are sufficiently bright so as to justify the premium price paid. An interesting question that we will return to later is whether this goodwill is a permanent asset or one that will diminish in value over time.

You may at this point quite reasonably view this list of assets as somewhat incomplete. Are these really the key assets owned by the company? What about, say, customer loyalty, or the aggregate scientific expertise in the development department, or the "culture" of the corporation, the set of beliefs and processes that differentiate it from its competitors? Indeed they are not physical assets, but they are almost surely the corporate strengths that the CEO trumpets in her annual report to shareholders. They must be important—and they are—but they are devilishly difficult to value, as we will explore further in Chapter 3. Moreover, they are bound up in and with the employees of the corporation. These employees go home every night and do not have to come back to work the next day; in no sense, therefore, does the corporation "own" these vital assets. Their value to the corporation lies in their future productivity but the financial statements record only history.

You might jump to the conclusion that an increase in total assets is a good sign, suggesting strong performance. Not so. Performance is best judged by analysis of the income statement. An increase in total assets may result simply from inefficient use of assets or it may accompany corporate growth in revenues. Moreover, if revenue growth could be accomplished with no addition to assets, so much the better.

Liabilities

Now turn to the other side of the balance sheet. The definition of **Current Liabilities** parallels that of current assets: liabilities that must be discharged within the next 12 months. I use the term "discharged" rather than "paid," because indeed some of the liabilities require certain performance by the corporation rather than the payment of cash. For example, down payments (advance payments) provided by customers remain a liability—typically a current liability—until the company provides to the customer the goods or services for which the down payment was made.

Do not fret about some of the unfamiliar nomenclature that appears in the current liability section; we will get to those definitions in Chapters 3 and 4.

Note that amounts due more than a year into the future are classified as long-term liabilities. An example would be installments on a 5-year term loan; those installments due within the next 12 months are classified as current and the remainder as long term.

As with the asset side of the balance sheet you might wonder if this enumeration of the corporation's obligations is complete. For example, the corporation has issued purchase orders for goods and services to be received in future accounting periods; in due course the goods or services will arrive and the accounts payable will be recognized. In the meantime, we might (but do not) record the obligation on the liability side and balance it with an asset labeled something like "right to receive" goods or services. Why bother? Accounting is complicated enough without making extra work for the accountants. Let us just wait until the goods or services—say, inventory—show up and then the value of both the asset Inventory and the liability Accounts Payable will be increased by the same dollar amount. Similarly, employment contracts might be valued equally as a liability and a "right to future services," an asset, but why bother—and moreover no employee is an indentured servant!

Assume that within the account **Accrued Liabilities** are some wages and salaries owed to employees when in the future they actually take vacation leave that they have earned. How is this obligation different from obligations associated with issued and outstanding purchase orders? The difference is that the corporation has already received the "value" for those future vacation wages in the form of work done by the employees; employees "earn" (we generally say "accrue") vacation when they work, not when they actually take the vacation time off. The same is true of the corporation's obligation to pay pensions to current and former employees.

How about lawsuits that the company will have to defend (lawsuits are frequent but unpredictable in our litigious society)? If we do not know who the plaintiffs will be or what wrongdoing they will allege, we do not have much basis for valuing this lingering liability. But suppose the corporation manufactures champagne corks; it knows from experience that champagne drinkers have a predilection to injure their eyes by exploding corks into them, and then to sue the manufacturer; the corporation may decide that these lawsuits are so frequent and predictable that valuing this liability on the balance sheet makes sense. Warranty obligations also fall into this category.

ONCE AGAIN: WHAT IS OWNERS' EQUITY?

The fundamental accounting equation tells us that owners' equity—net worth—is the difference between the value of the corporation's assets and the value of its liabilities. The funds to invest in the corporation's assets came in part from creditors and in part from its shareholders. Note that creditors consist of more than just those outside entities that have formal loan agreements with the corporation; they include also vendors (trade suppliers who provide goods and services on a "credit" or open-account basis); employees to whom wages are owed for work already accomplished; taxing agencies for taxes owed on past profits but not yet due and payable; and so forth. Assets can neither exceed nor fall short of the funds available to the corporation (any funds in excess of current requirements will appear as cash or cash equivalents, a current asset). Thus, the equality inherent in the accounting equation must hold.

Thus, owners' equity equals the amount of funds made available to the corporation by its shareholders. These funds come in two forms: (a) money invested by shareholders when they purchase newly issued common stock and (b) funds earned over the life of the corporation— profits—not paid out as dividends to shareholders but rather retained by the corporation for reinvestment on behalf of the shareholders. Owners' equity, therefore, is the sum of **Invested Capital** and **Retained Earnings**:

$$\text{Assets} - \text{Liabilities} = \text{Invested Capital} + \text{Retained Earnings}$$

A quick word of caution: do not think of owners' equity as a pool of cash. At all times a corporation's owners' equity has been used, just as borrowed funds are used as soon as they are borrowed—to pay for the assets that the corporation owns, including that most important asset, cash. Remember that the "liabilities and owners' equity" side of the balance sheet is invested in the "assets" recorded on the other side.

Could a corporation have negative retained earnings? Yes, and many do, particularly start-up companies whose accumulated losses (negative earnings) exceed their accumulated profits. Could a corporation have negative Owners' Equity; that is, could its liabilities exceed its assets? Yes, but such a corporation is in a precarious position because it has some very nervous creditors who have, you can see, loaned the corporation more than the value of its assets. Accordingly, negative owners' equity is not generally a condition that can be sustained; legal bankruptcy is likely to ensue.

Please remember that new owners' equity investment occurs only when the corporation issues, and shareholders purchase, new common stock shares. Thereafter any shareholder may trade with other investors, buying shares or selling shares that he or she owns. But this trading does not involve the corporation whose shares are being traded. Each common share represents a fractional ownership interest in the corporation and these interests are traded on the New York Stock Exchange, NASDAQ, and many other exchanges around the world.

Occasionally, companies purchase their own shares in the market place (or in private transactions with individual major shareholders) at or near the then prevailing market prices; when they do so, they typically retire the shares—in effect they simply tear up the shares.

Why buy one's own shares? Sometime these share repurchases are prompted by a desire to buy out a dissenting shareholder. Other times the corporation's board of directors decides simply that the prevailing market price represents a good bargain, a good place to invest some of the corporation's excess cash. Finally, such repurchases tend to push up the company's share price both (a) because the corporation is adding to purchase demand for outstanding shares with no change in supply of shares, and (b) because with the retirement of shares, the each remaining outstanding share represents a slightly higher fractional interest in the company and thus should be worth more.

Will these fractional interests—shares—trade at a price equivalent to their value reflected on the corporate balance sheet? That is, will the **book value** per share equal the **market value** per share. Typically, no: the market value may be higher or lower than the book value. Book value is easily calculated: total owners' equity divided by the number of common shares issued and outstanding. How is market value per share determined? By the free interaction of demand by investors for the shares and the supply of shares made available by their current owners. Investors seek to purchase shares that they think will provide future returns—in dividends and market price appreciation—more than commensurate with the current price. Sellers have the opposite expectation about the future. Note the key word in these last two sentences: future. Investors are buying into the future of the company. Book value is an historic valuation, the result of all of the past transactions and events in which the corporation has been involved. Unsurprisingly, "hot" stocks tend to have market prices greater than their book values, while "dog" stocks (such as the U.S. auto industry in the early years of the twenty-first century) frequently have market prices below their book values.

HOW IS THE EQUALITY OF THE BALANCE SHEET MAINTAINED?

The balance sheet balances. If it does not, an accounting error has occurred and needs to be corrected. To maintain that equality requires that every transaction recorded on the balance sheet have two equal and opposite elements, or entries as they are called. Accounting systems around the world are therefore referred to as **double-entry** systems.

Here are some simple examples that involve just the balance sheet (we will get to the income statement in the next chapter):

- Receive new inventory and pay cash for it. The value of the asset Inventory increases by the same amount as the value of the asset Cash decreases, and the equality is maintained. One asset is swapped for another.

- Receive new inventory purchased on credit. The value of the asset Inventory increases by the same amount as the value of the liability Accounts Payable, and the equality is maintained. Both assets and liabilities grow, but owners' equity is unaffected.

- Receive cash from a customer who purchased services last month on credit. The value of the asset Cash increases by the same amount that the value of the asset Accounts Receivable declines. Again, one asset is swapped for another.
- Pay an amount owed to a vendor for services received last month. The value of the asset Cash decreases by the same amount that the value of the liability Accounts Payable decreases. Both assets and liabilities decline but owners' equity remains unchanged.
- Borrow cash from the bank. The value of the asset Cash increases and the value of the liability Bank Loan Payable increases by the same amount. Again, owners' equity is unchanged, but the totals of both assets and liabilities grow.
- Pay vacation wages to an employee who is now taking vacation leave that she accrued over the past six months. Both the asset Cash and the liability Vacation wages payable decrease.

All of these transactions affected the financial position of the company—by rearranging assets and liabilities—but none affected the financial performance of the company. That is, the corporation's profit was not affected.

SINGLE-ENTRY VERSUS DOUBLE-ENTRY BOOKKEEPING

Is **single-entry bookkeeping** ever used? Yes, you and I both probably use single-entry bookkeeping to keep account of our personal finances. We increase our cash (bank balance) when we receive cash, regardless of when we might have earned the salary or wage, and we decrease cash when we write a check, even when the check is used to pay off a credit card balance, a liability, that may be several months old. Once in a while we are asked for a personal balance sheet, for example when we apply for a mortgage or other bank loan. Although we do not maintain a balance sheet on a day-to-day basis, we can rather easily construct one:

a) We think through and write down the value of the major assets that we own (a car, a house, securities, savings accounts, perhaps expensive jewelry, a cellar of fine wines or antique furniture; we probably ignore the value of our clothes, food in the pantry, and old furniture that would have minimal resale value).

b) We think through and write down the value of our outstanding obligations such as mortgages, credit card balances, auto loans (but we probably ignore property taxes that will be due late in the year, and this month's utility bills that sit unpaid on our desk).

c) The difference between the total values of our assets and liabilities thus determines our personal net worth.

Some very simple businesses can also appropriately use single-entry bookkeeping: a shoeshine stand, a lawn-mowing service, perhaps even a law office with just a single lawyer. Nevertheless, the huge majority of commerce conducted around the world is conducted by enterprises that use double-entry accounting systems.

ACCOUNTING PERIODS

The paragraphs above frequently referred to "this month" or "previous months." What is the significance of the temporal period "month"? Many, but not all, accounting systems use month as the relevant interim **accounting period**. That is, management decides that it wants to see complete financial statements—balance sheet and income statement and probably other financial reports as well—on a monthly basis. If so, the accounting department is expected to separate the transactions and events associated with a particular month from all those associated with earlier or later accounting periods.

Constructing financial statements is a time-intensive activity. Most companies do not want to do it every day; not enough changes in the span of a single day. On the other hand, reviewing financial statements only at the end of a year is too infrequent; management wants interim information to guide its decisions. Thus, typically a balance sheet is constructed at the end of each interim financial period—say, a month—and an income statement is constructed for that month.

Other definitions of interim accounting periods are sometimes used; the retailing industry often uses 13 four-week periods to constitute a financial year in order that each interim period have the same number of weekends. For more or less the same reasons, others use four 13-week periods to constitute a financial year, rather than using calendar quarters.

But in essentially all cases the key accounting period is the year, referred to as the **fiscal year**. Fiscal years can start on any day, but typically they commence on the first day of a month. The choice of fiscal year is often a function of the organization's activity; educational institutions generally start their fiscal years on July 1 or September 1 and retailers avoid fiscal years starting on January 1, right in the middle of their busiest season.

DEBITS AND CREDITS

We have made it this far without uttering the intimidating words, **Debit** and **Credit**. We need not dread or avoid these words; they are simply accounting jargon for the condition of individual accounts and for increases and decreases to those accounts—rather like sailors use port and starboard for left and right! But, unlike port and starboard, debit and credit seem to have emotional connotations to many of us. We tend to think of credit as "good"—we like credits to our bank accounts or credits to our charge cards when we return defective merchandise—and we dislike debits—the annoying $1.50 debit to your bank account every time you use an ATM machine! Well, try to shake yourself loose from these emotional contents of the words.

Asset accounts carry debit balances. Liabilities and owners' equity accounts carry credit balances. Total debit balances must *always* equal total credit balances. Logically, then, an increase in an asset account is shown by a debit entry and a decrease by a credit entry. And, the opposite is true for liabilities and owners' equity accounts: an increase is shown by a credit entry and a decrease by a debit entry. In each of the six examples of transactions shown above, the debit entry equals the credit entry.

By convention, accountants never enter negative numbers in accounting records. That is, a decrease in an asset is not effected by a "negative debit entry" but rather by a credit entry.

So, in the accounting system we have both debit balances and debit entries, and both credit balances and credit entries. Before and after each transaction is recorded, the total debit balances equal the total credit balances and therefore, of course, the complete recording of a transaction must have a debit entry (or entries) equal to the credit entry (or entries).

We will expand these definitions of debit and credit in the next chapter to include those accounts that appear on the income statement.

COULD THE BALANCE SHEET SUFFICE AS THE ONLY FINANCIAL STATEMENT?

Could we get along with just a balance sheet and chuck the income statement? You do not expect me to argue for that proposition, but let us explore its ramifications.

Consider a simple merchandising company. How would we record a sales transaction? The asset Inventory is credited (decreased) by the value of the merchandise—presumably the cost to the company of acquiring the item(s) from suppliers—now being handed over or shipped to the customer; the asset Cash (or Accounts Receivable if this is a credit sale) is debited (increased) by the selling value of the merchandise, a value that is, or presumably should be, higher than the merchandise's acquisition cost. So now the debit and credit entries do not balance, the balance sheet is out of balance, and we need another entry—more credit. That credit is profit (actually, gross profit or margin on this one sales transaction) and so adds to owners' equity (retained earnings). When the company pays the monthly salary to a sales clerk, the Cash account is credited (decreased) but no other asset or liability account is affected; this salary payment is an expense and thus reduces owners' equity (retained earnings).

Following this logic, we could simply record sales and expenses on the balance sheet—still using double entries—and ignore the income statement. The owners' equity account would go up (receive a credit entry) as revenues are achieved and be reduced (a debit entry) whenever expenses are incurred. At the end of the accounting period, profit or loss for the period would be the amount by which retained earnings increased or fell.

But, without an income statement we will forego lots of information that is really essential for financial statement readers to have. We will move on to the income statement in Chapter 2.

EXAMPLE: FEDERATED DEPARTMENT STORES BALANCE SHEET

Exhibit 1-2 shows the balance sheet of Federated Department Stores Inc. (lightly edited to simplify) at the end of its 2004 fiscal year. Federated owns many department store groups, including such well-known names as Macy's and Bloomingdale's, has executive offices in Cincinnati, Ohio and New York City, employs more than 110,000, and operates 491 department stores and 460 bridal and formalwear stores in 46 states. It defines its 52-week fiscal year as comprising four 13-week periods and its end-of-year balance sheet is dated January 29, 2005. As a merchandising company, Federated has a somewhat more straightforward set of financial statements than would a manufacturer.

Note two interesting alerts in the title of this exhibit. First, the word "Consolidated" alerts us that Federated adds together (rolls up or consolidates) the balance sheets of all of its subsidiaries and certain of its affiliates (financing, logistics, data processing, and other subsidiaries). Second, note the parenthetic: given Federated's size, there is little point in showing the values down to the last penny; although I assure you that Federated's accountants know the value to the last penny! Instead, for presentation purposes, values are rounded to the nearest million dollars. A smaller company might report values to the nearest thousand dollars.

What characteristics of this large merchandising company are evident on its balance sheet? It has substantial investments in merchandise inventory. It has an inventory of supplies as well, but presumably because they are modest in value, they have been combined with the prepaid expenses. Federated's retail store properties dominate its property and equipment account (fixed assets)—over $6 billion—almost twice the value of its merchandise inventory. The footnotes to these statements tell us that about $1 billion of this is land, $4.2 billion is buildings, including leasehold improvements, and $4.6 billion is fixtures and equipment (store furnishings primarily); these values are at original cost of these assets but accumulated depreciation on these assets totals $3.8 billion, netting to the $6.0 billion shown on the balance sheet. Federated's asset investments total nearly $15 billion—high for a company whose total revenues were a bit less than $16 billion (as we will see on its income statement in the next chapter). Federated's accounts receivable are about 22% of the year's revenue; while this may seem high, remember that the balance sheet reflects January 29 values, quite soon after the holiday busy season.

Note that liabilities exceed owners' equity: about 60% of the assets are financed with debt, and the remainder with shareholders' equity. Federated has both long-term and short-term

Exhibit 1.2: *Federated Department Stores, Inc. Consolidated Balance Sheet January 29, 2005 ($ millions)*

ASSETS

Current Assets	
Cash & cash equivalents	$868
Accounts receivable	3,418
Merchandise inventories	3,120
Supplies and prepaid expenses	104
Total current assets	7,510
Property & equipment (net)	6,018
Goodwill	260
Other intangible assets	378
Other assets	719
Total assets	$14,885

LIABILITIES AND SHAREHOLDERS' EQUITY

Current Liabilities	
Short-term debt	$1,242
Accounts payable & accrued liabilities	2,707
Income taxes payable	352
Total current liabilities	4,301
Long-term debt	2,637
Deferred income taxes	1,199
Other liabilities	581
Shareholders' Equity	
Invested capital	3,124
Retained earnings	4,365
Treasury stock	(1,322)
Total shareholders' equity	6,167
Total Liabilities and Shareholders' Equity	$14,885

debt. The portion of long-term debt that is due within the next 12 months is included in short-term debt.

Because the account Deferred Income Taxes appears on most U.S. corporate balance sheets these days, a word of explanation is in order. Tax laws are written to raise funds for the government, not to reflect or to drive sound accounting practices. Quite appropriately, companies, like individuals, seek as low taxable income as is permitted by law, so as to minimize their cash payments of income taxes. Typically, taxable income reported to the governmental taxing agencies is lower than that reported to shareholders and other audiences. If this sounds like keeping two sets of books, it is—legally, appropriately, and with full knowledge of the taxing agencies. But, the result of this "two sets of books" business is that very frequently the income tax expense reported on the income statement is more than the company will have to pay in the coming year. The actual amount owed is shown as a current liability and the balance is added to the deferred income tax balance. As you can see, this deferred income tax at Federated is not a trivial item: nearly $2 billion. For reasons that we do not need to go into now, deferred taxes can often be deferred for a very long time; thus, this $2 billion is in effect interest-free, long-term "borrowing" from the government for Federated.

Finally, look at the composition of Federated's owners' (shareholders') equity; unsurprisingly for a very mature company, its retained earnings are almost half against its invested capital—even after the company has paid out as dividends to its shareholders a meaningful portion of its profits (last year, about 15%). Incidentally, the balance sheet does not reveal the amount of those dividends because its purpose is solely to reflect the financial position of Federated at January 29, 2005; but the Cash Flow Statement discussed in Chapter 6 does provide this information.

The next-to-the-last line on the balance sheet, in the owners' equity section, shows the value of the so-called **Treasury Stock**. This represents what Federated paid for shares of its own common stock that it purchased in the market. Earlier, I suggested that companies in effect tear up such shares when they repurchase them; because the value shown here is a negative number, the accounting treatment that Federated uses amounts to the same thing. Incidentally, at first glance it might appear that Federated repurchased about 40% of its stock, since the value of Treasury Stock is about 40% of the value of Invested Capital. Not so; Federated paid a lot more for each share of common stock it repurchased than it received many years earlier when most of the common stock was issued and sold by the company.

A BALANCE SHEET REFLECTS THE NATURE OF THE BUSINESS

Would a balance sheet of another kind of business look similar to Federated's? Yes, the structure is the same. But, no, the relative importance of various assets and liabilities on the balance sheet may be quite different, driven by the different nature of the business the company conducts.

Compare in your mind a department store group such as Federated with an electric power utility such as Commonwealth Edison (CE) that generates and distributes electricity. Compared to Federated, CE has huge investments in fixed assets: generating plants and distribution facilities. Its inventory is relatively small: just fuel for its generating plants. Its long-term debt is high both because it needs the funds to invest in fixed assets and because the steady and predictable nature of its business (a regulated public utility) permits it to service this high debt with relatively little risk of default. CE's investment in accounts receivable (that is, the total owed to it by its customers) is modest because its customers pay their bills in a timely manner; you can imagine that CE has an effective way to assure prompt payment by its customers!

Think for a moment about commercial banks. What are their primary liabilities? Bank deposits that its customers (you, me, companies, and so forth) entrust to the bank; these deposits are our assets, but they are liabilities to the bank. And, its primary assets are loans: promises to pay executed by its borrowers. When an individual or company borrows from the bank, the borrower's liabilities increase while the bank's assets increase correspondingly.

Now compare a supermarket with a department store like Federated. At first glance, they may look much alike, but the supermarket's investment in assets, as a proportion of its sales, will be lower than the department store. Its inventory of food moves (turns over) more quickly—it had better, to assure freshness—and it owns essentially no customer accounts receivables (though its customers may well use Visa or other major credit cards). The supermarket need not hold a high cash balance (i.e., safety stock of cash) because its business is not seasonal (as the department store's is) and cash purchases by its customers are both steady and highly predictable.

Think about a manufacturer of commercial aircraft like Boeing. The in-process manufacturing time for large aircraft is necessarily long, and thus Boeing has high inventory values (but minimum finished goods inventories, as completed aircraft are immediately delivered to the customer). Given the sorry financial position of most large airlines, at least in this country in the early years of the twenty-first century, the aircraft manufacturers may have to provide generous financing terms to its airline purchasers, with resulting high values of accounts and notes receivable.

Some of these struggling airlines have negative retained earnings. In fact, in the entire 80-plus year history of the airline industry, cumulative losses have exceeded cumulative profits! (It is a wonder that new airlines continue to be formed!) Of course, many of these large U.S. airlines have declared bankruptcy in recent years. As interesting as the question is, this is not the time or place to speculate on whether or how these bankrupt airlines can rearrange both their finances and their operations to exit bankruptcy and remain solvent.

NEW TERMS

Accounting equation: assets = liabilities + owners' equity

Accounting period

Accounts payable (A/P)

Accounts receivable (A/R)

Accrued liabilities

Balance sheet

Book value

Credit entry and balance

Current assets

Current liabilities

Debit entry and balance

Double-entry accounting

Fiscal year

Fixed assets

Goodwill

Intangibles

Inventory

Invested capital

Liquidity

Market value

Net worth

Retained earnings

Shareholders' (stockholders') equity

Single-entry accounting

Treasury stock

CHAPTER 2

The Income Statement
Performance Not Position

As the balance sheet is a statement of financial position and not performance, the **Income Statement** is a statement of financial performance, not position. A company with a strong balance sheet could have weak profit performance for several years, although persistent weak performance will in time weaken its financial position. Another company might achieve handsome profits and growing sales revenue—that is, strong income statement performance—and at the same time be in weak financial condition as reported on its balance sheet.

The purpose of the income statement is to help us understand how the company is performing, how successful it is at generating revenues and profits, and what factors contribute to that strong or weak performance.

As every balance sheet is dated and represents a snapshot of financial position as of that date, the income statement specifies a period for which it is reporting the corporation's financial performance. The universal, key accounting period is a nominal year (in part because corporations must report taxable income annually); that year generally commences on the first of a calendar month, but not necessarily on January 1. The operations of the corporation are accumulated for the year, but are also generally reported monthly and/or quarterly (or for a 4-week period and/or a 13-week period).

Chapter 1 asserts that an accounting system could get along with only a balance sheet, never mind the income statement. Every sale would increase retained earnings (within the owners' equity section of the balance sheet) and every expense would reduce it. The change—increase or decrease—in retained earnings between two balance sheet dates equals the profit (or loss) for that time interval. So, what is the purpose of the income statement? It is simply—but importantly and usefully—an elaboration or amplification of the changes in retained earnings for the accounting period. Accordingly, the linkage between these two fundamental products of the accounting system, the balance sheet and the income statement, is straightforward: it is the profit; the final line of the income statement is the amount by which retained earnings grew or declined during that period.

Exhibit 2.1: *Typical Income Statement Structure*

	Dollars
Sales/Revenue	$100
Less: Cost of Goods Sold	60
Gross Profit (Gross Margin)	40
Operating Expenses	
Engineering and development	6
Selling and marketing	13
General and administrative	10
Subtotal	29
Operating Profit	11
Other Revenue and (Expense)	(2)
Profit Before Income Taxes	9
Taxes on Income	3
Net Income	$6

The income statement is often referred to as the **Statement of Operations** or the **Profit and Loss (or P&L) Statement**. For simplicity, we will stick with the name income statement.

STRUCTURE OF THE INCOME STATEMENT

Exhibit 2-1 shows the typical, if somewhat stylized, structure of an income statement. The top line (or lines) always shows the **Sales** or, an equivalent name, **Revenue** for the period for which the income statement is reported. Think of the dollar amounts in Exhibit 2-1 as index numbers or percentages, with total revenue equal to 100% and net income in this example equal to 6% of revenue. But remember that these numbers will vary greatly for different types of businesses, as discussed further below.

Sales include only those goods or services delivered during the current accounting period. The corporation may have received other orders during the period, orders for goods or services to be delivered or provided in future accounting periods; those orders will not appear as sales/revenue until those future periods.

Cost of Goods Sold for the period includes the cost only of those goods or services for which revenue is recorded in this period. Thus, cost of goods sold does not include the cost of all merchandise received in this period (in a merchandising enterprise) or the cost of all goods produced (in a manufacturing enterprise). Rather, cost of goods sold is **matched** to the revenue in order that the **Gross Profit** (or **Gross Margin**, equivalent terms) will have useful meaning to the readers of financial statements. Gross profit (margin) indicates the aggregate amount by which sales values exceeded acquisition costs (of merchandise in a retail environment) or production costs (in a manufacturing company)—clearly useful data!

But, of course, any corporation—merchandising, service, or manufacturing—incurs other expenses in addition to those reflected in cost of goods sold. These are referred to as **Operating Expenses** and they too are "matched"—but matched to the accounting period rather than to sales/revenue. That is, accountants must include *all* operating expenses relevant to the accounting period, but exclude those that pertain to earlier or later periods. For example, the company's facilities rental agreement may call for quarterly payments, but if the relevant accounting period is a month rather than a quarter, the accountant needs to include only the equivalent of one month's rent. Chapter 4 discusses further the techniques used to assure proper matching of revenues and expenses.

Exhibit 2-1 shows the typical categories of operating expenses for a manufacturing company: engineering and development; sales and marketing; general and administrative. Obviously, a retail enterprise that incurs few or no engineering/development expenses will omit that category. Such a company might want to show a separate line item for advertising and promotion distinct from other sales and marketing expenses. In short, defining relevant categories of operating expenses is generally left to the company and is a function of the nature of its business.

Gross profit less operating expenses equals **Operating Profit**, as indicated in Exhibit 2-1. The obvious next question is: what would constitute **Nonoperating Revenue and Expense** or, as it is labeled in Exhibit 2-1, "other revenue and expense"? Included here are revenues and expenses that are tangential, peripheral, or incidental to the business—that is, not mainstream, given the definition of the corporation's primary business. Examples of such other revenue include rental revenue derived from the excess space that the company happens to own, or interest revenue derived from cash-equivalent securities in which the company has invested its excess cash. The primary nonoperating expense at most companies is interest expense, i.e., interest on its borrowings. The amount of interest expense is a function of how the corporation chooses to *finance* its activities, not how it operates its business. Exhibit 2-1 shows that this hypothetical company had greater nonoperating expenses than nonoperating revenues and thus the net of the two is shown in parentheses, that is, as an expense.

In countries like the United States, where the primary form of corporate taxation is taxes on profit, the next line on the income statement is the Profit Before Income Taxes, followed by Income Tax Expense.

The bottom line here could be labeled Profit, but its formal name is **Net Income—**logical, since this is the Income Statement. Exhibit 2-1 shows Net Income of 6% of revenues, an amount that is probably slightly above the median figure for U.S. corporations. But this percentage varies enormously both by industry and by individual company. Pharmaceutical and software companies often report after-tax net income of 25% of revenues; U.S. auto companies are thrilled if they report any positive net income; and struggling companies frequently report negative net income, that is, a loss.

AN INCOME STATEMENT REFLECTS THE NATURE OF THE BUSINESS

As with a company's balance sheet, much about the nature of the company's business is reflected in its income statement.

A supermarket chain can expect only a very modest gross margin percentage. The supermarket can still be adequately or even handsomely profitable, despite a low gross margin, both because it typically generates very high sales volumes in relation to total investments (total assets) and because its operating expenses (clerks' salaries, store rental, utilities, and so forth) are quite modest.

In contrast, a high-tech instrument manufacturer had better be able to generate a very significant gross margin. The high-tech manufacturer typically must incur high engineering, development, and selling expenses to maintain its technology edge and to convince customers to purchase complicated state-of-the-art instruments; it must generate a high gross margin so that it is still left with reasonable net income after these high operating expenses.

An electric public utility also must generate a high gross margin in order that it can meet the high expenses associated with extensive and expensive fixed assets and pay the interest charges on its large borrowings and still have a reasonable net income.

Some service companies report a low gross margin because most of its expenditures involve salaries to professionals and these salaries are included in its cost of goods sold (more logically referred to as cost of services rendered). Banks must earn sufficient spread between the interest revenue they receive on loans and interest expenses they incur in attracting deposits (which provide the funds that are loaned) to cover their operating expenses.

WHY ARE DIVIDEND PAYMENTS NOT CONSIDERED AN EXPENSE?

Interest on loans represents return to the lender and expense to the borrower. **Dividends** are income/revenue to the shareholder who receives them, but they are not recorded as an expense of the corporation paying them. Why are these two forms of return on capital—interest and dividends—treated differently? The short answer is: convention. Many "rules" of accounting derive simply from convention. But a more complete explanation is that shareholders are the ultimate owners of the business, the group for whom the business is being operated. Whether the profits of the business are paid out in dividends or retained and reinvested in the business—or, frequently, some of each—the profits are benefiting the shareholders.

Early in this chapter we commented that Profit—Net Income—adds to Retained Earnings. True. And dividends paid are recorded as a debit to Retained Earnings (part of owners' equity) and a credit to Cash (both are reductions). Thus, for the accounting period the *net* addition to owners' equity is net income less dividends paid.

DEBITS AND CREDITS

It is time to return to the subject of debits and credits (port and starboard!) and relate these terms to the income statement. Recall that revenues/sales increase, in effect, owners' equity while expenses decrease it. In the last chapter, we noted that increases in owners' equity (as well as liabilities) are reflected by credit entries (and thus typically carry a credit balance) and decreases in owners' equity are reflected by debit entries. It follows, then, that sales are recorded with credit entries and expenses with debit entries. Sales balances in the accounting records carry credit balances and expenses carry debit balances. Incidentally, you may be happier with the emotional meanings of the words debit and credit in the context of the income statement: you are more enthusiastic about credit entries than debit entries—i.e., you prefer revenues to expenses—but all that is irrelevant!

To recap then,

- debit entries record increases in assets or expenses;
- credit entries record increases in liabilities, owners' equity and sales/revenue.

And, we do not make negative entries; rather

- decreases in assets or expenses are recorded with credit entries;
- decreases in liabilities, owners' equity, and sales/revenue are recorded with debit entries.

Accordingly, asset and expense accounts typically have debit balances and liability, owners' equity and sales/revenue accounts typically have credit balances. And remember that the sum of *all* the credit balances must equal the sum of *all* debit balances. To maintain that equality, we must continue to follow double-entry bookkeeping.

Some examples, now involving both balance sheet accounts and income statement accounts:

- Checks are drawn to pay office salaries; this transaction is recorded by a credit entry to Cash (decreasing an asset) and a debit entry to Office Salaries Expense (increasing an expense account).

- The telephone bill, covering the current month's charges, is received but will not be paid until next month; the debit entry increases the Telephone Expense account, and the credit entry increases Accounts Payable.

- A check is received (and deposited) for this month's interest on a loan the corporation made to a company employee; the debit entry increases the asset Cash and the credit entry increases Other Income (a nonoperating revenue).

- A check is drawn to pay the janitorial service firm for last month's services. Note that the amount charged by the janitorial service firm should have appeared among *last* month's expenses; it is not an expense of the current month. Accordingly, this amount should now reside in the balance of the Accounts Payable account. The entries are, then, a debit entry to decrease Accounts Payable and a credit entry to decrease Cash. Note that this month's income statement (and net income) are unaffected by this transaction.

- A credit sale is made to a customer. The credit entry increases Sales/Revenue and the debit entry increases Accounts Receivable. But this transaction is more complex: the corporation also reduced its inventory by the value of merchandise shipped to (or handed over to) the customer. Thus another pair of entries is required: debit to increase Cost of Goods Sold and credit to decrease Inventory.

Many people have trouble getting their minds around the fact that both assets and expenses carry debit balances, and that both liabilities and revenues carry credit balances. In fact, one of the common dilemmas accountants face is whether a certain debit entry should be made to an expense account or an asset account.

To illustrate, suppose a corporation purchases for $400 cash miscellaneous office supplies (pads of paper, pencils, staples, etc.). It does not buy just enough supplies for the current month, but enough to last, say, 3 months. The transaction is recorded by a credit entry to Cash (reducing the asset cash) but where should the debit entry go, to Office Supplies Inventory (an asset) or Office Supplies Expense (an expense)? A debit entry to Office Supplies Expense will reduce this month's profitability; a debit entry to the asset will not. Either is theoretically OK, and since the amount involved is minor, we are probably indifferent as to where the debit entry is made. As a practical matter, the company will probably next month buy other office supplies in quantities sufficient to last several months (binders, paper clips, ballpoint pens). Since the amount spent each month on office supplies is small, why not take the easy way out and record the debit entry in the expense account?

You probably realize that if the corporation had purchased the office supplies on open account (i.e., for credit), the credit entry would not be to Cash but rather to Accounts Payable.

But suppose the expenditure was substantially larger, for example, re-roofing the factory. Re-roofing occurs very infrequently, and the new roof will extend the life of the building. If we treat this expenditure as an expense—by, say, debiting the account Maintenance Expense—we negatively distort reported profit for the month when the re-roofing occurred. It makes more sense to place the debit entry into a fixed asset account; the accountants can then adjust the value of this fixed asset account over the life of the new roof.

GENERAL LEDGER AND CHART OF ACCOUNTS

Chapters 1 and 2 have used liberally the word "account." What is an "account" and how many are there? The accounting system needs to categorize or classify the various assets owned, liabilities outstanding, revenues achieved, and expenses incurred. That classification is accomplished by having separate accounts for each category of asset, liability, revenue, and expense that we want to be able to examine separately. How many accounts? As many as you need, and no more! The more accounts, the more detail available for analysis. But, also, the more accounts, the more accounting time required, and the greater the possibility that analysts will get lost in the trees and fail to see the forest. The availability of high-speed computers has tended to push accountants in the direction of excessive detail, so much data that useful information gets lost. Nevertheless, hundreds or even thousands of accounts, depending upon the size and complexity of the corporation's activities, may be required to capture the level of detail desired. Of course, each of these accounts does not appear in the corporation's published financial statements; for presentation purposes accounts are combined appropriately so that the financial statements are of manageable length.

Accounts exist in what is called the **General Ledger**, the fundamental accounting books of the corporation. You need to be aware of this term, but we need not here delve into the format and operation of general ledgers. The listing of all accounts used in an accounting system is referred to as the **Chart of Accounts**. You can probably imagine what a typical chart of accounts looks like—a kind of roadmap of the accounting system, but pretty boring reading!

You need to be aware that the Chart of Accounts does *not* contain several accounts that you might expect it to, for example, Gross Margin, Operating Profit, Net Income, Total Current Assets, Total Assets, and so forth. These amounts—the first three on the income statement and the next two on the balance sheet—are all calculated or derived as the accountant assembles the financial statements. For example, Gross Margin is the difference between total sales/revenue and total cost of goods sold, both of which are included in the general ledger. Similarly, by summing all the sales/revenue account balances accumulated throughout the period and subtracting the sum of all the expense account balances, we derive net income for the period.

Exhibit 2.2: *Federated Department Stores Consolidated Income Statement Year Ended January 29, 2005*

	$ Millions
Net Sales	$15,630
Cost of Sales	9,297
Gross Margin	6,333
Selling, General and Administrative Expenses	4,933
Operating Income	1,400
Interest expense	(299)
Interest income	15
Income Before Taxes on Income	1,116
Income tax expense	(427)
Net Income	$689

EXAMPLE: FEDERATED DEPARTMENT STORES INCOME STATEMENT

Continuing the financial statement example shown in Chapter 1, Exhibit 2-2 shows the Federated Department Stores Income Statement (again, lightly edited) for the year ended January 29, 2005. Recall that Federated is a very large merchandising operation; the dollar amounts in Exhibit 2-2 are in millions.

The titles on the income statement are largely self-explanatory, but two clarifications are in order. The top line is "Net" sales, that is, gross sales less returns (which are typically not minor for department stores). Undoubtedly Federated executives track closely the pattern of sales returns, but the company has chosen not to provide that information to its external audiences. Second, note the use of the title Cost of Sales, a perfectly acceptable alternative to the Cost of Goods Sold label that we used earlier.

Federated generates very substantial annual revenues—almost $16 billion in 2004—and enjoys a gross margin in excess of 40%; that is, its cost of merchandise is about 60% of what it sells the merchandise for. Note the classification of the operating expenses that Federated uses: selling, promotion, general and administrative expenses are all lumped together.

If we are trying to analyze Federated in detail, and perhaps compare its financial performance (income statement) and position (balance sheet) with its primary competitors, we might wish for more detail, particularly a separation of its selling expenses and its administrative expenses. We saw on Federated's balance sheet (Exhibit 1-2) that the company borrows substantial amounts, and thus we are not surprised to find that Federated's interest expense is a hefty amount, $299 million. At year-end, the sum of Federated's interest bearing debt, short term and long term, was just under $4 billion. We can estimate that the average interest rate on this debt must have been about (299 divided by 4,000) 7.5%. Federated's interest income was only $15 million; this seems low in light of a $868 million balance in Cash and Cash Equivalents at year-end (see Exhibit 1-2). If an average of $500 million of the Cash was invested in interest bearing securities throughout the year, the $15 million of interest income implies a 3% interest rate on these securities; one would of course expect this rate to be well below the rate that Federated pays on its borrowing. We see that Federated's effective income tax rate (federal, state, and local income taxes) is just under 40%; that is, its tax expense is 40% of pretax profit. Finally, note that Federated's net income in 2004 was about 4.4% of net sales. It would be interesting to compare that earnings rate to that of other retailers.

Much can be learned about Federated by relating the financial data in Exhibit 2-2 (the income statement) to certain balance sheet data shown in Exhibit 1-2. While these relationships are explored in detail in Chapter 7, we can hint at some of them here. Note that Federated generates less than $16 billion of revenue with a total asset investment of almost $15 billion. What does that say about the productivity of its investments? In the previous paragraph we looked at the relationship between net income and sales, but if we relate net income to total owners' equity, we get an idea of how profitable the company is in relation to the amount that shareholders' have entrusted to it. Finally, note that both Merchandise Inventory on the balance sheet and Cost of Sales on the income statement are expressed in terms of the cost to Federated of acquiring the merchandise that it sells; a comparison of those two amounts can tell us a good deal about how "fast" on average Federated turns over (moves) its inventory. But very fast turnover is not necessarily a positive sign; chances are higher with fast turnover that a merchandiser will be out of stock of items the customer seeks. But it is also true that slow turnover can be a problem; particularly, with fashion merchandise that goes quickly out of style.

FOOTNOTES TO FINANCIAL STATEMENTS

Exhibits 1-2 and 2-2 show Federated's balance sheet and income statement. These, however, do not comprise Federated's full financial reporting for the fiscal year ended January 29, 2005. Another formal statement, the Cash Flow Statement, will be taken up in Chapter 6. But an essential and very illuminating part of the financial reporting is the extensive **Notes to the Financial Statements**, actually an integral part of the statements themselves. In Federated's 2004 annual report, these notes, set in quite small type, take up 35 pages of text.

So, here is a good place to admonish you readers: *always read the financial footnotes*, even though unfortunately they are often obscure and legalistic. Traditionally the first note provides detail on the company's key accounting policies; Federated devotes seven pages of small type just to enumerating its accounting policies. Companies have some leeway as to accounting policies they follow, and you need to be aware of its policies as you review the financial statements. Later in the notes, details regarding the company's borrowing agreements with banks and other lenders, and a schedule of required loan repayments in the years ahead, are spelled out. The status of any legal proceedings in which the company is involved as either plaintiff or defendant is disclosed. If the corporation has acquired other companies, the terms of the acquisition(s) are the subject of another note. Compensation of top executives and the company's use of incentive programs, including stock options, are also included. Details of the tax expense on the income statement and the tax liabilities recorded on the balance sheet are also available in the notes. Extensive detail is available in the notes about the company's pension obligations, as well as other retirement benefits it provides present and retired employees.

NEW TERMS

Chart of accounts

Cost of goods sold (Cost of sales)

Dividends

General ledger

Gross margin

Gross profit

Income statement

Matching (match)

Net income

Nonoperating revenues and expenses

Notes to financial statements

Operating expenses

Operating profit

Profit and loss statement (P&L)

Revenue

Sales

Statement of operations

CHAPTER 3

Valuation
Where Disagreements Arise

Thus far, we have encountered little disagreement or controversy about how financial statements are assembled or what their structure should look like. Other countries may use formats or structures a bit different than those used in the United States, and goodness knows different nomenclatures are used (for example, the British use the term "turnover" where we use "sales/revenue"), but they all

- present in the balance sheet Assets = Liabilities + Owners' Equity, or a derivative thereof, and
- link the income statement and the balance sheet by adding profit earned to retained earnings.

In this chapter and the next—focused, respectively, on valuation and timing—we will see that sensible people can disagree both as to the values of transactions and events and as to timing: when they should be reflected in the financial statements.

DEFINITION OF ACCOUNTING

Although we will not delve deeply into the mechanics of accounting, we can usefully spend a moment on the definition of the process:

Accounting is the process of *observing, measuring, recording, classifying, and summarizing* the changes occurring in an entity, expressed in monetary terms, and *interpreting* the resulting information.

In this definition, the action words are italicized. This chapter and the next focus on *observing* and *measuring* (valuing). We have noted that accounting data are *classified* according to the accounts listed in the chart of accounts. We will minimize the discussion of the *recording* process. Chapter 7 focuses on *interpreting* financial statements.

It is worth repeating that accounting records can only deal with those transactions, events, and conditions that (a) can be expressed in monetary terms, and (b) have already occurred—are history! Take care also to define the entity to which the accounting pertains. This is not difficult for corporations, but, for example, note that if a CEO buys common shares in her company from a third party, this transaction does not impact the corporate entity. And, if an individual owns 100% of his small business, it can be difficult to separate the accounting for the owner from accounting for the business.

VALUATION: COST, MARKET, OR TIME ADJUSTED

This chapter focuses on valuation, a somewhat more complicated business than it may appear. Three basic approaches are used: **cost, market, and time adjusted**. Before we explore each, let me jump to the conclusion. All three methods are used and useful, but cost values predominate. This domination is not because "cost" wins out on the basis of theory or elegance or even accuracy. Rather "cost" wins on practicality. As intellectually appealing as market and time-adjusted valuations are, in a majority of cases they are simply too unreliable and difficult to apply.

But before moving to these three valuation methods, it is worth reemphasizing that if we get the values of assets and liabilities absolutely correct—admittedly, an unattainable utopia—we will necessarily and perforce accurately value owners' equity since it is simply the difference between assets and liabilities. And, if we get owners' equity valued absolutely correctly, we automatically have an accurate valuation of net income.

The flip side of that truism is the key: if we overvalue assets (but accurately value liabilities), we will overstate retained earnings and thus overstate net income. And, if we understate the value of liabilities (but accurately value assets), we will also overstate retained earnings and thus net income. Now we begin to see the challenge, and glimpse the temptation for unscrupulous managers to misstate earnings. We will revisit this challenge frequently throughout the remainder of this book.

Valuation at Cost

To simplify the discussion, examples in Chapters 1 and 2 used assumed values. When Federated bought merchandise from its suppliers, we assumed that it should be valued in Inventory on the balance sheet—and eventually as Cost of Sales on the income statement—at what Federated paid for it, its cost. And, we do not have any trouble with valuing the sale of this inventory at what the customer paid for it (the cost to the customer). Fair enough.

But suppose we were accounting for a manufacturing company that transforms raw materials and components purchased from its vendors into a finished product that it sells to its customers. The original raw material and components can be valued at cost, but in-process inventory as it moves through the manufacturing process increases in value. We need the procedures of cost accounting—see Chapter 8—to measure in monetary terms this increase in value.

In general, transactions are valued at cost—cost to the acquirer: cost to Federated when it buys merchandise; cost to Federated's customer upon sale of the merchandise. The "other side" of the accounting entry—accounts payable and accounts receivable—utilizes the same value, of course, in order that the equality of the accounting equation is maintained.

When a corporation buys a truck, a computer, a desk, a generator, or a piece of production equipment, that new fixed asset is valued at its cost. Incidentally, we typically include in that cost ancillary charges for delivery, tax, and installation required to place the fixed asset in a useful position and condition.

Current assets and current liabilities are reasonably straightforward to evaluate since, by definition, they are very liquid. As we move down the list of assets and liabilities to less liquid items, we encounter substantially more valuation challenge.

Valuation at Market

Occasions arise when valuation at original cost is inappropriate—and still other occasions when we may be tempted to value at *market* rather than at original cost, but need to resist the temptation. Here are some examples.

Suppose Federated negotiates a very good deal on certain new merchandise—a price well below "the market price" offered to its competitors. Should it value that merchandise at prevailing market rather than at its bargain price? If so, how would Federated account for the transaction? Debit the asset Inventory at the market price and credit Accounts Payable at the negotiated bargain price? The entries are out of balance and the other necessary credit (the difference between the two prices) will have the effect of improving Federated's profit for the period. Well, that result does not sound unreasonable: clever purchasing improved the performance at Federated; why not recognize that happy fact?

Sorry, accounting for the bargain purchase in this manner violates accepted practice. Accountants are constrained to use *cost* rather than *market* in valuing the inventory. The entire profit on the purchase-and-subsequent-sale of the merchandise—the full gross margin—is recognized at one moment: when the sales transaction is effected.

You might argue that, if the merchandise is sold on credit, the full gross margin is not realized until the customer pays her bill, since after all some accounts receivable inevitably prove uncollectible. Well, that argument, too, is trumped by accepted practice. After all, if the creditworthiness of the customer is very weak, Federated probably will not extend credit to the customer. Nevertheless, in exceptional cases it may be reasonable to postpone recognizing any gross margin on sales to very poor credit risks until the seller has the cash in hand.

Change the example to a laptop computer manufacturer that maintains sizeable inventories of integrated circuits. The legendary Moore's law predicts that the market value of these integrated circuits will decline during the time they languish in inventory. Should market prices, rather than cost prices, be used to value this inventory of circuits? Yes, but before we discuss why, consider the opposite condition: the manufacturer also maintains an inventory of precious metals, the market prices for which increase subsequent to acquisition but while they are still in inventory. How about using market prices here, rather than cost prices? No.

How do we reconcile this apparent contradiction? The rule is: in valuing inventory use the **lower of cost value or market value**. This rule will result in a conservatively low valuation of inventory, early recognition of expense associated with holding inventory, and a postponement in the recognition of any profit arising from holding inventory. The decline in inventory value of the integrated circuits is reflected as soon as it is discovered. The fortuitous gain in value of the precious metals is postponed until those metals find their way into laptop computers sold to customers and their values thus appear in cost of good sold.

Consider the valuation of fixed assets during the years that they are used; that is, well after they were acquired. How about using market values? There are two problems here:

- Valuing any asset that does not have a ready second-hand market, for example specialized machine tools and intellectual property, can be very problematic.

- Market values are quite irrelevant if the owner has no intention of selling the asset; the benefit of owning fixed assets derives from their use, not their resale.

In Chapter 4 we will discuss methods used to reflect the declining values of fixed assets over the course of their useful lives.

But some assets do indeed have reliable and ready markets, for example actively traded securities. Suppose a corporation invests some of its excess cash in fixed-income securities (bonds and notes) issued by government agencies or other corporations. The market price of these securities moves up and down as prevailing interest rates move down and up. Unlike fixed assets, these securities (assets) are held by the corporation for the purpose of being sold (liquidated) when the corporation needs the cash. Still other corporations, such as mutual funds and Berkshire Hathaway (the company led by the legendary Warren Buffett), hold listed equity securities (common stocks) in unrelated companies. In the case of such assets, the rule is to **"mark to market"**; that is, value them at their market prices, reflecting any "paper" gains or losses during the period as revenues or expenses of that period.

Time-Adjusted Valuation

A third valuation method I refer to as *time adjusted*. Its use is primarily in the valuation of assets and liabilities that consist of a stream of future cash receipts or payments. Consider a quick example. The corporation advances a newly hired key executive $200,000 to assist her with relocation expenses, with the understanding that she will repay the advance at the rate of $40,000 per year for 5 years with no interest charges. How should the corporation account for this advance? How about a credit to the asset Cash and a debit to another asset called something like loan receivable? Without doubt the credit to Cash is correct, but does the corporation really have a "loan" asset appropriately valued at $200,000? No; its value is less because the corporation is giving up the opportunity to invest the $200,000 in another productive way. Assuming that the corporation can earn 10% per year on these other productive investments, we can time-adjust these future payments at the 10% rate and determine that the loan receivable has an initial value of about $151,000. If the asset Loan Receivable is valued at $151,000, where does the other $49,000 debit go (so that the two debit entries match the single $200,000 credit to Cash)? To an expense account; in effect, the corporation has provided a "hiring bonus" to the new employee and that expense should be recognized in the accounting period when the "advance" was funded.

Many readers will recognize that the value of the advance or loan is the present value of the future stream of cash flows calculated by the well-known and accepted techniques referred to as Discounted Cash Flow, or DCF. Widely available computer programs and interest tables facilitate the calculation of present values.

In fact, many (perhaps all) securities are valued in the marketplace by these techniques. Take, for example, a 10-year, $1,000 face-value bond with a stated interest rate (coupon rate) of 7%. If the prevailing rate of interest for a bond of its creditworthiness was 7% at the time the bond was initially sold (by the issuing government agency or corporation), the bond sold for $1,000. Now suppose 2 years later prevailing interest rates for bonds of this creditworthiness have increased to 8.5%. The market price of the bond will adjust: the future stream of interest and principal payments will be discounted (time adjusted) at an 8.5% rate, and its present value will be less than $1,000. This lower market price will, in effect, offset the fact that the bond's coupon rate is not competitive in the current interest-rate environment. Bond markets work in exactly this way: when prevailing rates increase, bond market prices decline, and vice versa. Clever investors are eager to buy bonds when they anticipate that prevailing interest rates will decline in the months or years ahead.

Note, too, that a bond's market value will decline when the issuer's creditworthiness declines; that price decline has the effect of increasing the effective interest rate (called the yield) on the bond to compensate bondholders for accepting the additional risk associated with deteriorating credit quality.

Arguably, the market value of common stocks is also the time-adjusted (discounted) value of all of the future cash flows associated with owning the stock: future dividends and price appreciation. These future cash flows are, in effect, time adjusted at an interest rate that investors believe appropriate in light of the riskiness (or risk profile) of the corporation. Differences of opinion among investors as to (a) future cash flows and (b) riskiness and therefore appropriate discount rate can cause one investor to be eager to sell a particular common stock while another investor is eager to buy it.

A simple numerical example may help clarify this valuation method. What is the value of $1,000 in cash flow—inflow or outflow, that is, asset or liability—per year for 10 years? You certainly would not pay $10,000 for these promised cash inflows over the next 10 years. What you *would* be willing to pay depends on what returns you could earn on alternate investments (assuming, again, similar risk) and these alternate returns determine the discount rate that you would apply in calculating what you would pay. Here are the maximum prices you would pay at alternative discount rates:

Discount Rate	Time-adjusted Value
6%	$7,360
8%	6,710
10%	6,144
12%	5,650
15%	5,019

Thus, if the discount rate appropriate for you is 15%, this 10-year promissory note has a value to you of only one-half its nominal $10,000 value.

The time-adjusted valuation method has a great deal of appeal—at least to me! My enthusiasm for the method is tempered by the difficulty of estimating future cash flows, to say nothing of appropriate discount rates, for most assets and liabilities. Nevertheless, the time-adjusted valuation is both appropriate and widely used for financial assets and liabilities. In addition to the examples already cited—loans and receivables—the technique is used to value employee pension obligations and other retirement benefits that involve corporate cash outflows running many years into the future. (Interestingly, many large corporations are not currently fully valuing their pension liabilities.) Indeed, any liability that will not be fully discharged until years into the future should be adjusted for time; similarly, any asset that promises the corporate owner cash benefits years into the future should be time adjusted.

ESTIMATING AND APPROXIMATING VALUES

You may have the impression that values recorded on balance sheets and income statements are entirely objective, derived from hard evidence. Many are, but not all. When it would be either too costly or too delaying to insist on perfect valuation knowledge, the accountant needs to estimate or approximate values. The benefits of financial information decline rapidly with time; that is, readers of financial statements are content to trade good, but not perfect, accuracy in timely financial statements for precision in much delayed statements. These readers must make decisions and they want the most up-to-date information they can get, even if some estimated values must be used.

Here are some examples:

- A corporation is sued for $10 million; after careful consideration of the merits of the suit, management and the lawyers believe the suit can be settled for about $1 million. In recording this liability the $1 million estimate should suffice, since it may be years before the exact cost of the lawsuit is known. But valuing this liability at zero would be misleading.

- Another corporation decides to terminate a product line and sell the equipment used exclusively to manufacture that product. Once that decision is made, the equipment is no longer a typical fixed asset, but rather an asset available for sale. The company must estimate what it will realize in the sale and revalue the equipment accordingly.

- A company determines that its inventory of part T36 is obsolete; it will be sold at a fraction of original cost (the value currently reflected in inventory), but that exact fraction is unknown. An estimate based upon the best knowledge available will suffice for revaluing this obsolete inventory.

- When recording the increased pension obligation for the current accounting period, the corporation cannot know precisely retirees' life expectancies and future compensation. In fact, the valuations of all pension obligations are predicated on estimates; actuaries are the professionals who provide the corporation with well-informed estimates.

- A maintenance firm subcontracts a major activity on a certain maintenance job completed during this accounting period. The subcontractor has not yet submitted an invoice for it; the maintenance firm should estimate the subcontractor's charges and include that amount in the Cost of Services account this period.

- An airline values its obligation to frequent-flyer customers who have "miles earned" but not yet redeemed. The airline has no difficulty measuring the number of miles outstanding, but valuing them is quite another matter. Airlines know that a high percentage of these miles will never be redeemed. By the end of 2004, passengers had earned nearly two trillion miles but redeemed less than one-half trillion. If a "free" ticket is used on a flight that otherwise would have empty seats, the incremental cost of the "free" passenger is minor: some costs associated with booking the flight, checking the passenger in, handling baggage, a bit more fuel, but not much else. On the other hand, if the "free" passenger bumps a paying customer off a full flight, the opportunity cost of the "miles" is high. No doubt the frequent-flyer obligation should be recorded, but valuing that obligation requires many assumptions. These assumptions will become more accurate as the airline gains experience with the program.

What do you do when your estimates (sometimes even guesstimates must be utilized) prove incorrect—and, to a greater or lesser extent, that will be the usual situation? Except when the estimate is both profoundly wrong and the impact on profit is very large, we do not go back to change the accounting of a prior period. If the lawsuit requires $1.1 million to settle, $100,000 more than the estimate, that $100,000 is recorded as an expense in the period when the lawsuit is settled. If the subcontractor's invoice turns out to be a few percent higher or lower than the estimate, no big deal! The difference adds to or reduces cost of services rendered in the future period. If a company overestimates the "charge" associated with closing a manufacturing plant, the amount of the overestimate appears as a negative expense (credit to an expense account) in that future period when the overestimate becomes apparent. Incidentally, companies facing the tough decision of closing a plant or division have a tendency to overestimate the attendant costs—to get the bad news behind them. The estimated charge is generally large, widely recognized as nonrecurring, and thus typically ignored in assessing performance of the remaining on-going business. The later "correction" to the estimate is likely to be small and serves to improve future profits.

But, return to the lawsuit example and assume that the settlement finally cost the full $10 million, a rather staggering amount for the small-company defendant. In such a circumstance, it may indeed be appropriate to restate financial results for the previous period when the lawsuit's financial impact was woefully underestimated. In recent years, a disquieting number of financial restatements have been undertaken to correct earlier distorted or biased reporting. Companies with sound and conservative financial reporting restate very rarely.

ACCOUNTING FOR CURRENCY FLUCTUATIONS

In these days of globalization and worldwide corporate activities, companies increasingly face financial reporting issues arising from doing business in many different currencies. Large companies have subsidiaries around the world, each keeping its accounting records in the local currency. When the parent company constructs consolidated financial statements (recall that Federated's statements were "consolidated"), all amounts denominated in foreign currencies must be translated into—that is, valued using—the home currency. This translation is not difficult, as rates of exchange between currencies are well established because billions of dollars of currency transactions occur daily. However, when a company does business in more than one currency, it is exposed to currency gains and losses.

Suppose Corporation T, based in the United States, sells goods to a Japanese customer valued at 40,000 Japanese yen. At the time of the sale the yen/dollar exchange rate is 100 yen per dollar. Thus T records the sale (credit entry) and accounts receivable (debit entry) at $400. Now suppose that, before the Japanese customer pays, the yen/dollar exchange rate moves to 120 yen per dollar. The amount owed by the customer has not changed: it remains 40,000 yen. Has the value of T's account receivable changed? Yes; T must revalue it at (40,000 divided by 120) $333, and recognize a **currency translation loss** of $67. At the end of every accounting period Corporation T must revalue assets and liabilities denominated in currencies other than the U.S. dollar.

Of course, Corporation T can "hedge" against currency gains and losses by entering into compensating foreign exchange contracts. Note that T can also hedge by borrowing as many yen as it lends; if T *owed* the same amount of yen in the form of accounts payable as it *owned* in accounts receivable, when the exchange rate moved from 100/1 to 120/1, its loss on the A/R would be exactly offset by its gain on the A/P.

SOME VALUATION RULES

We can summarize most of the major points of this chapter by enumerating a few accounting rules that must be followed in valuing assets and liabilities:

1. *Cost* values dominate, although (as with most rules) there are exceptions where *market* or *time-adjusted* values are appropriate.

2. Gross margin is *realized* when the sales transaction occurs, no portion before and no portion later. We call this rule the **realization** concept.

3. Inventories are valued at the lower of cost or market value.

4. Valuations should be *conservative*. When in doubt, choose the valuation that results in the lower statement of profit in the current period. The stereotype of the *conservative* accountant undoubtedly derives from this important principle or rule. This rule suggests that when considering alternative valuations, the accountant should lean in the direction of understating the values of assets and revenues, and lean in the opposite direction in valuing liabilities and expenses. Note that this rule does *not* say understate assets and revenues and overstate liabilities and expenses.

5. The test for valuations is that they be *feasible, verifiable, timely,* and *free from bias.* If the valuation method is elegant (as I find the time-adjusted method to be) but in the circumstances is not feasible or cannot be arrived at in timely fashion, forget it! The valuation must have some basis in fact so it can be verified; it must not be arrived at by

some biased whimsy. This is the rule that pushes relentlessly in the direction of using cost rather than market or time-adjusted values.

6. Financial assets available for sale at any time by the corporation should be "marked to market" values.

7. Certain values shown in balance sheet accounts (and therefore also reflected in income statements) must frequently be estimated. With a heavy emphasis on speedy financial reporting, accountants are willing to trade precision for timeliness.

8. One more valuation rule, not yet discussed, deserves mention. We make the **going-concern assumption** when valuing assets and liabilities. That is, we assume that the entity for which we are accounting will remain in its current business. When and if this assumption becomes no longer valid, valuations will alter significantly. Fixed assets will be valued at what the company can realize for them in the second-hand market. Bad debt experience is likely to worsen as the company will have no on-going relationship with customers. Inventory may have to be sold at distress prices. At the same time, liabilities will tend to persist at their former valuations, with the result that owners' equity may suffer a significant decline.

NEW TERMS

Accounting (formal definition)	Lower of cost or market
Conservatism	Mark to market
Cost valuation	Market valuation
Currency translation gain/(loss)	Realization
Going-concern assumption	Time-adjusted valuation

CHAPTER 4

Timing
The Corollary to Valuation

Chapter 3 focused on methods to value assets, liabilities, revenues, and expenses. This chapter focuses on the issue of timing—when should those values be recorded? Recall from Chapter 2 (the income statement discussion) that matching is all-important: matching cost of goods sold to sales/revenues, and matching operating expenses to the accounting period.

One of the valuation rules presented in the previous chapter was *conservatism*. This rule applies even more in determining the proper timing of revenues and expenses. The rule can be stated slightly differently here: when in doubt as to timing, lean in the direction of delaying the recognition of revenue but lean in the direction of the early recording of expenses. This rule will result in the conservative statement of net income (profit) and therefore owners' equity.

Chapter 3 reviewed six examples where the use of estimated values was appropriate. Each of these examples also illustrates the importance of timing:

- Once the $10 million lawsuit is filed and the corporation determines that it is unlikely to prevail in court, the corporation needs to recognize the probable cost of settling the suit: debit entry to an expense account and credit entry to a liability account.

- The termination of a product line typically involves costs. When should those expenses be realized? At the time the termination decision is made. Recall how often you see in the media that a large manufacturing company "has taken a major charge"— that is, recorded a large expense—associated with its decision to close a plant. Severance pay for dismissed workers is typically the major portion of that expense. Recording this severance obligation calls for a debit to an expense account and a credit to a liability account when the decision to close is made, not in the future periods when the severance payments are made. (Note that those future severance payments will be recorded by a debit to the liability account and a credit to the asset Cash, with no impact on profits in those future periods.)

- Similarly, obsolete inventory should be written off—debit to an expense account and credit to the inventory account—at the time it is determined to be obsolete. Of course, the inventory may not be worthless; if management believes it can be sold at a much-reduced price, only the difference between that estimated price and the original cost of the obsolete inventory should be written off.

- When should the expenses associated with retirement pensions be recorded? Employees "earn" pensions during their working lives and "receive" pension payments during their retirement. Accordingly, the corporation accumulates pension obligations as the employees work. Thus, pension expenses should be recorded pro rata, period by period, just as salaries and wages are recorded period by period. Surprisingly, current pension accounting rules give employers wide latitude in when they recognize these obligations. For some companies these so-called "underfunded" pension obligations are enormous. Recognized or not, these obligations are real—and troublesome!

- The cost of subcontracted work for which the invoice has not yet been received must be matched against the revenue on the completed maintenance job. And, if the maintenance job was completed during this accounting period, the associated revenue should be reported in this period. Thus, the estimated subcontractor charges must be included in this period's cost of goods sold (here more appropriately called cost of services delivered).

- The liability of frequent flyer miles must be accrued (accumulated or built up) accounting period by accounting period as passengers earn the miles. This liability should be revalued as the airline develops experience regarding the usage of the miles (or changes the rules for redeeming miles). As this revaluation leads to increases or decreases in the liability, it also results in decreases or increases in profits.

Here are 12 other "timing" issues addressed in this chapter:

1. Wages are earned in this accounting period but paid in a future accounting period.
2. Interest expense is paid quarterly, not monthly.
3. Insurance premiums are paid annually, in advance.
4. A 25% down payment is made on a new fixed asset to be received 3 months hence.
5. The company anticipates that it will need to spend some amount in the future to effect warranty repairs of its products sold in this period.
6. The company receives the telephone bill covering charges from the middle of the last month to the middle of this month.
7. The office supplies inventory was depleted somewhat in this month.
8. Some accounts receivable from customers will prove to be uncollectible.

9. Inventories tend to "shrink" over time from theft, loss, and obsolescence.

10. Fixed assets decline in value through their useful lives.

11. Purchases are made on which the vendor offers a discount for prompt payment of the bill.

12. Inventory contains materials or items purchased at different times and at different prices.

MATCHING, ACCRUAL, AND DEFERRAL

The majority of timing questions arise because of the need to match: match sales and expenses to the accounting period and match cost of goods sold to revenues/sales. The other major cause is that cash is frequently paid out or received in one accounting period but the expense or revenue is properly recognized in an earlier or a later accounting period.

We will use the first five examples listed above to illustrate *accruals* and *deferrals* that are necessitated by the matching principle.

Suppose a company pays its employees on the 10th of the following month for work performed in the current month. Obviously salary and wage expenses need to be recorded this month, even though cash payments will not be made until next month. We say that salary and wages are **accrued** and reflected as a corporate liability at the end of the month. In addition, suppose that the employees earn the right to one vacation day for each month worked—in Europe it would be at least two days per month! The company needs to accrue this vacation liability by debiting salary and wage expense this month for one day of compensation and crediting that amount to the liability account entitled something like Accrued Vacation Salary and Wages.

The second example is similar: an appropriate amount of interest expense needs to be recognized each month, although cash outflow occurs only at the end of the quarter. Thus, the interest obligation is accrued each month with a debit entry to Interest Expense and a credit entry to Accrued Interest Payable, a liability account. At the end of the quarter, this liability is discharged: debit to the Accrued Interest Payable account and a credit entry to Cash. Failure to make these entries understates expenses during the first two months of each quarter and overstates interest expense in the final month when the interest payment is actually made.

In other cases, such as the third example above, the cash flows out before the accounting period when the expense should be recognized. For obvious reasons, insurance premiums must be paid in advance. If the insurance protection runs a year into the future, a portion of that premium (presumably 1/12th) should appear as an expense in each month. That is what the matching principle demands. When the cash is paid out, the company is purchasing an asset: insurance protection for a year. The appropriate entries, then, are a debit of the full premium amount to an asset account typically called **Prepaid Expenses** and a credit to the Cash account. This prepaid account is then drawn down month by month over the coming year by 12 equal entries: debits to Insurance Expense and corresponding credits to the Prepaid Expense account, so that at the end of the year the Prepaid Expense account for insurance has a zero balance.

The fourth example also involves an early flow of cash. If a company pays in advance (a down payment, sometimes called "earnest money") $100,000 on a $400,000 specialized piece of equipment being customized for its use, the $100,000 payment does indeed "purchase" something of value, but not the fixed asset itself. Thus, the debit entry is again to Prepaid Expenses and the credit entry is to Cash. When the new equipment arrives, this prepaid expense is reduced to zero. By the way, if the name Prepaid Expenses does not sound quite right to you, you are quite free to utilize another more descriptive asset name like Down Payments—but be sure that the account is in the chart of accounts, or take the necessary action to add it.

Not infrequently, a seller has an on-going responsibility to service a customer without receiving any additional postsale compensation. This pricing arrangement is often called "bundled pricing": the price of the future services is bundled into the price of the goods. Example 5 above presents such a situation: Company T sells an instrument or machine for which it provides the customer a warranty. Another example: software companies typically promise updates or telephonic consulting to customers who buy its software products. Some equipment manufacturers "bundle" the cost of installation and of training customer personnel with the price of the equipment itself. Surely you can come up with still other examples.

The point is that the cost of this future servicing should be matched appropriately. Two approaches are utilized, both requiring accruals/deferrals. Assume that Company T expects that after-sale service (warranty, training, and so forth) will on average cost 2% of the sales value of the instrument.

One approach involves simply deferring part of the sales revenue to a future period when Company T will incur the associated service expenses. Conservatively, Company T might defer 3% of the bundled sales revenue, thus deferring part of the gross margin to the future period(s). This approach involves a debit of the full sale amount, of course, to Accounts Receivable (assuming that the customer has agreed to pay the "bundled" amount upfront) that is balanced by two credit entries: 97% of the sale is credited to Sales in the current period (where it will be matched by an appropriate cost of goods sold value) and the 3% balance to a liability account called something like **Deferred Income**. That account name appropriately signals that more income is coming, but it will be recognized in a future accounting period. In these future periods, part or all of the deferred income is accounted for by a debit to Deferred Income (thus reducing that liability) and a credit to a revenue account, perhaps Service Revenue. Now the cost of performing the service is properly matched to these increments of service revenue and part of the "bundled" gross profit is realized in the future period(s).

It may trouble you to have deferred income—which sounds like a desirable thing—treated as a liability. But indeed the seller is obligated to the customer for this after-sale service for which the company has paid. It really is a liability!

The other approach is to include in Cost of Goods Sold the estimated cost of performing the after-sale service, together with the cost of producing the equipment or software, and match that combined total to 100% of the "bundled" revenue in the period when the goods are delivered to the customer. Accordingly, 2% of the bundled price is included in Cost of Goods Sold and that amount is credited to a liability account called **Accrued Expenses**. Note that fulfilling this future servicing obligations will not reduce net income in those future periods; the entries will be a credit to Cash (or perhaps Wages Payable) and a debit to the Accrued Expenses liability account.

MATERIALITY

Examples 6 and 7 in the above list permit an illustration of the principle of **materiality**. This principle states simply that accountants should focus attention on meaningful accounting entries and not worry unduly if they are not precisely correct either in amount or in timing. If you will, close enough is good enough!

Perhaps, theoretically, telephone and other recurring monthly invoices should be split between accounting periods if indeed their charges relate to more than a single one. But telephone charges do not vary much month to month— that is, the differences are immaterial— and thus so long as one monthly telephone bill (or electric bill or waste collection bill) gets included in each monthly accounting period, both accounts payable and net income will not be materially misstated.

As to office supplies, the company could keep close control over this inventory, and thus be able to value accurately what was added to and withdrawn from the inventory in any accounting period. Alternatively, it may decide that the value of this inventory is immaterial and moreover the company tends to buy some amount of office supplies every month: pencils and pens this month, computer paper next month. Why not just treat these purchases as expenses at the time they are made and ignore valuing this minor asset on the balance sheet? You might do the same with janitorial supplies—but not if you are accounting for a janitorial service company; for such a company, the value of the janitorial supplies inventory is probably material.

Thus, materiality must be judged on the basis of the size of the company and the nature of its business. A small company may feel that the purchase of a laptop computer for the marketing vice president is a material expenditure, and thus it should be treated as a fixed asset, while another larger company that (a) buys laptop computers frequently and (b) believes that the useful life of a laptop computer is very short, may decide to "expense" (debit to an expense account and credit to cash or accounts payable) each laptop when it is acquired.

ALLOWANCES; CONTRA ACCOUNTS

Examples 8 and 9 suggest the need to adjust two asset accounts—accounts receivable and inventory—for statistically predictable future events that cannot be precisely identified and measured at this time. Almost any company that sells "on credit" to its customers knows that occasionally a customer will be unable to pay; or more correctly, the company will be unable to force him or her to pay! And every retailer, and many other kinds of businesses as well, know that their inventories will "shrink"—a polite euphemism for stolen or lost.

The argument for adjusting these values as reported on the balance sheet is that "bad debts" and "inventory shrinkage" are current costs associated with effecting credit sales and holding inventory. These costs should appear as expenses in the period when the credit sales are made and the inventory is maintained, not later when we discover exactly what customer(s) will not pay or what inventory items disappeared.

Suppose Company A sells beauty products to beauty parlors on credit. Beauty parlors get started and go broke with great regularity; thus, beauty parlors can be considered to be relatively poor credit risks. If Company A has been in this business for a few years, it probably can discern a pattern of bad debts—about what percentage of credit sales prove ultimately to be uncollectible, let us say 4%. Another company in another kind of business may have negligible credit risk, and therefore no need to adjust its accounts receivable. But Company A does have that need. Of course, A does not know which customers will not pay; if it did, it would decline to sell to those customers. But in an accounting period when it sells $400,000 of beauty products to beauty parlors, it estimates statistically that 4%, or $16,000, will never be collected. These bad debts are a cost associated with this period's credit sales and, logically, this expense should be matched to this period.

This matching is accomplished by a debit entry to an expense account called perhaps Bad Debt Expense, but where does the offsetting credit entry go? Not to the Accounts Receivable account, because we do not yet know which receivables will be uncollectible. Rather, the credit goes to what we call a **contra account**—an account that is "contra" or offsetting to the Accounts Receivable account. This contra account often goes by the euphemistic name of **Allowance for Doubtful Accounts**. (I am still looking for the accounting system that calls this account somewhat more accurately, if pejoratively, "Allowance for Deadbeats.") In the chart of accounts the Allowance for Doubtful Accounts generally appears right after Accounts Receivable despite the fact that it carries a credit balance. Think of it as a negative asset because it adjusts the accounts receivable balance. On the balance sheet the two accounts are combined into

"Accounts Receivable (Net)"—gross accounts receivable reduced by the amount of the allowance account.

Now, what happens when Company A determines that Customer Mu has gone bankrupt leaving a $6,000 account unpaid? Well, of course, the accounts receivable account must be reduced by $6,000—a credit entry—but the $6,000 debit entry is to Allowance for Doubtful Accounts, also reducing that account. The key here is that the $6,000 is *not* an expense of the accounting period when A finally finds that Mu will not or cannot pay. Profits of that period are unaffected by that discovery. The expense (and the corresponding credit to increase the Allowance account) was matched to the period when the credit sale was made.

Any asset can have a contra account to adjust its value. The obvious other example is inventory; see example 9 above.

ACCOUNTING FOR FIXED ASSETS

The primary challenge in accounting for a fixed asset is to find a logical, rational way to spread the cost of the fixed asset—more precisely, the difference between the asset's original cost and its estimated salvage value at the end of its useful life—over that asset's useful life. That, in fact, is the definition of **depreciation**. Note that it is *not* depreciation's objective to reflect year by year the market value of the fixed asset.

Suppose Company R buys a fleet of forklift trucks today for $145,000, including the cost of delivery. Company R estimates that it will replace these forklifts in 8 years, and at that time will be able to sell them for (they will have a salvage value of) $25,000. Pause for a moment and consider whether these are conservative estimates, as that term is used in accounting. The shorter the assumed life and the lower the estimated salvage value, the higher the year-by-year depreciation expense and therefore the more conservative the estimates. Estimating a 2-year life and zero salvage value for forklift trucks would be just silly, far too conservative. These estimates seem reasonable.

If the difference between original cost and estimated salvage value ($120,000) is divided evenly over the 8-year life, annual depreciation on this fleet will be $15,000. This is referred to as **straight-line depreciation**. In accounting for depreciation, another contra account is used, **Allowance for Depreciation**; the original cost of the asset remains in the primary asset account, Fixed Asset. Using this example, the debit of $15,000 is to Depreciation Expense and the credit entry is to Allowance for Depreciation. The so-called **book value** of this fixed asset at the end of 2 years is ($145,000 less $30,000) $115,000. At the end of its 8-year life the book value is of course the estimated salvage value.

A reasonable argument can be made for the use of **accelerated depreciation**, recognizing higher depreciation expenses in the early years of the asset's life (when it is new, most productive, and relatively maintenance free) and lower depreciation expenses when the asset is old and requires more maintenance. (Incidentally, accelerated depreciation also comes closer to reflecting the declining pattern of the asset's market value, but that is not a relevant argument since reflecting market values is not depreciation's objective!)

Many forms of acceleration are possible, but a common one is the so-called **double-declining-balance depreciation**. The rate of depreciation is double the straight-line rate—in this case 2/8 or 1/4—but this rate is applied to the declining book value. Thus, the depreciation expense and the book value for the first 3 years are (this method calls for ignoring the salvage value in calculating annual depreciation in the early years):

Year	Depreciation Expense	End-Of-Year Book Value
1	$36,250	$108,750
2	$27,187	$ 81,563
3	$20,391	$ 61,172

Obviously, accelerated depreciation is more conservative than straight-line depreciation.

Company R's published balance sheet at the end of year 2 will probably carry a single-line item entitled "Fixed Assets (Net)," included in which will be these forklifts valued at $81,563, the difference between the $145,000 original cost, in the primary fixed asset account, and accumulated depreciation of $63,437, in the contra account.

Assume that the original assumptions about this fleet of forklifts do not play out; new fuel-efficient models become attractive by the end of year 3 and Company R sells the fleet for $52,000 at that time. At the time of sale, the realized price is below the $61,172 book value; the company occurs a loss of $9,172. This sale is not recorded as normal revenue on the top line of the income statement, since R is not a forklift dealer and thus is not in the business of buying and selling forklifts. This is an unusual event, and thus recorded as follows:

Debit	Cash (or Accounts Receivable)	$52,000
Debit	Allowance for Depreciation	$83,828
Debit	Loss on Sale of Fixed Assets	$ 9,172
Credit	Fixed Assets	$145,000

This set of entries deserves a little explanation—in addition to the reminder that the sum of the debit entries equals the credit entry! Once Company R sells these forklifts, the asset values pertaining to them must be removed: the $145,000 in the Fixed Asset account and the $83,828 in the Allowance for Depreciation contra account. Those removals are accomplished with the second and fourth of the four entries. What kind of account is Loss on Sale Fixed Assets (generally labeled Gain/Loss on Sale of Fixed Assets)? It is part of the Nonoperating Income and Expenses that appears just below the Operating Profit line on the income statement.

Valuation of intangibles follows a similar pattern to the valuation of fixed assets. The anticipated decline in the value of the intangible (e.g., a patent) over its life is prorated across that life. Nomenclature differs slightly: intangibles are **amortized**, while fixed assets are depreciated.

DEPRECIATION AND INCOME TAXES

This is an appropriate time to raise the matter of income taxes. As you can imagine, the taxing authorities (the Internal Revenue Service in the United States) have plenty of rules about how companies account for depreciation when figuring their income tax obligations.

As mentioned earlier, taxing authorities' rules are established to raise revenue for their governments. In addition, tax rules are sometimes set to accomplish other objectives, most specifically to stimulate (and occasionally to dampen down) the economy. Economic activity is stimulated by increases in capital equipment expenditures. More liberal depreciation policies—that is, allowing greater depreciation early in the assets' lives—help that stimulation. Some governments go the whole way and permit companies to write off as a current expense for income tax purposes the full price of some or all new fixed assets.

Well and good, but, to repeat, none of this has anything to do with sound and appropriate accounting for fixed assets. For tax reporting purposes companies will depreciate their fixed assets as aggressively as possible: accelerated depreciation over short useful lives. Note that these depreciation expenses do not use up cash; they are a so-called noncash expense (more on this in Chapter 6 during the discussion of Cash Flow Statements). However, very often lower depreciation expenses are used in published financial statements, for example in corporate annual reports. The difference between these two accounting treatments—tax versus so-called book—give rise to deferred income taxes such as we saw on Federated's balance sheet.

Depreciation (and amortization) expense is often referred to as a tax shield: these expenses reduce taxable income and therefore the amount of cash that has to be expended to pay taxes, but, to repeat, do not consume cash. Of course, this tax shield is only useful if the company makes profits and pays taxes. So-called tax losses can be both carried backward—to reclaim income taxes paid in recent years—and carried forward to offset income taxes that would otherwise be due in the next several years. But some companies—think airlines—have not made profits in recent years and may not for years to come, and yet they require heavy investments in fixed assets. The tax shield associated with accelerated depreciation on these fixed assets does the airlines no good. Thus, the airlines typically lease their equipment from financial companies who can utilize these tax shields.

One more word on income taxes. The "loss on sale of fixed assets" at Company R (see above) is a deductible expense in calculating R's income taxes. Correspondingly, a gain would be taxed.

ACCOUNTING FOR CASH DISCOUNTS ON PURCHASES AND SALES

Numbers 11 and 12 on the list of examples above illustrate that accountants have some freedom in choosing accounting policies. These are interesting, but frankly do not typically have major impacts on financial statements.

To encourage speedy payment by their customers, many sellers offer cash discounts for prompt payment. Each of seller and buyer then has the choice to account for these discounts on either a **gross** or **net** basis. Suppose Company V is particularly eager to receive prompt payment from customers (perhaps because V is perpetually short of cash or because its customers, such as beauty parlors, need an extra inducement to pay promptly) and so offers it customers a 3% discount if the invoices are paid within 10 days. Suffice to say that this is a very attractive discount offer, and most customers will avail themselves of it. If so, should Company V record its sales net of the discount—at 97% of invoice value—given the expectation that customers will typically take the discount? Or, alternatively, should the sale be recorded at gross—100%—and the discount shown as an expense when (often in a future accounting period) the customer actually pays, taking the discount? Company V has the choice, but it should be obvious that the first alternative is the more conservative. If this "net" method is used and some customers do not take advantage of the cash discount, then the 3% discounts foregone by customers are treated as other income by Company V in the period when the customer pays 100%.

Accounting for cash discounts taken by buyers is the mirror image of the choice just outlined for sellers: treat the purchase as having a value net of the cash discount or record the discount as other income when the invoice is finally paid, perhaps in a subsequent period. Here the conservative approach is the second alternative, the "gross" method.

FIFO/LIFO INVENTORY VALUATION

When identical items in inventory are bought at different times for different prices, how does an accountant decide in what order these "costs" are moved from Inventory to Cost of Goods Sold? For example, suppose Company K makes the following purchases of part B472, all of which remain in inventory on April 10 when 75 units are withdrawn, assembled into K's primary product and shipped to a customer.

Date Purchased	Number Purchased	Cost Per Unit
February 8	100	$10.47
March 11	150	$11.10
April 4	50	$11.35

In calculating the cost of good sold for the product, what "cost value" should the accountant use for part B472? The company can choose any of the following three accounting policies:

- **FIFO (first-in, first-out).** Use the oldest cost, $10.47, for all 75 units.
- **LIFO (last-in, first-out).** Use the most recent costs, $11.35, for 50 units and $11.10 for the remaining 25 units.
- **Average cost.** Use the weighted average of all costs now in inventory (here $10.93) for all 75 units.

The choice among these options becomes more significant the higher the inflation rate. Persistent inflation, at various rates, characterizes developed as well as developing economies. Thus, accountants are more concerned with conditions of inflation, not deflation, when making this accounting policy choice. Another rule of accounting is **consistency**: once an accounting policy is selected, the company should stick with it for future accounting periods. Thus, it is not acceptable for a company to switch back and forth between LIFO and FIFO.

LIFO has the advantage of reflecting in COGS (cost of goods sold) the cost that most closely approximates replacement cost for part B472 (because it is the most recent cost). On the other hand, if the inventory of this item does not run down to zero periodically, its value may in time include some very "old costs"—and thus be an unrealistic value.

By contrast FIFO values inventory more realistically but COGS less realistically.

In inflationary times, LIFO states net income more conservatively and thus leads to lower taxes. An oddity of the U.S. income tax law provides that companies may not use LIFO for tax reporting unless they also use it for financial reporting. Why? Simply because Congress wrote the tax law that way! Incidentally, generally the first note in the Notes to the Financial Statements will tell you what policy the company is following in valuing its inventory.

Note that best practices in the *physical* flows of inventory follow FIFO: use up the old stuff first. But that does not preclude using LIFO or average cost for *monetary* flows of inventory.

RECURRING AND NONRECURRING ADJUSTING ENTRIES

Most accounting entries are triggered by one of four kinds of transactions:

- Sales of goods or services.
- Receipt of goods or services, including the service of employees.
- Payment of cash to discharge obligations, including salaries and wages.
- Receipt of cash from customers.

Of course, some other transactions also must be recorded: funds may be borrowed or repaid; shares of common stock may be sold or purchased; fixed assets may be bought or sold.

In addition, other events occur and conditions change that affect valuations and thus must be reflected in the accounting records, even though they did not involve transactions. That is, valuations must be adjusted to reflect these events or altered conditions.

Many of these so-called **adjusting entries** are called for at the end of each accounting period; that is, they are recurring entries. We have already looked at several:

- Depreciation expenses for the accounting period must be recorded.

- Prepaid expenses (such as prepaid insurance) must be adjusted to reflect a portion as expenses of the current period.

- Allowance for Doubtful Accounts, Allowance for Obsolete Inventory, Warranty reserves are adjusted.

- Salaries and wages are accrued for work performed for which payment has not yet been made.

- Certain assets and liabilities denominated in foreign currencies are revalued using current exchange rates.

In addition, managers and accountants must search out other appropriate valuation adjustments that do not occur on a regular basis. These nonrecurring adjustments are required to value accurately assets and liabilities. And remember, if we value the assets and liabilities properly, we will necessarily also properly value net income. Here are some examples of possible nonrecurring entries:

- The corporation's board of directors declares dividends to be paid in the next accounting period.

- The corporation announces an employee layoff that creates an obligation for severance payments to be made in future periods.

- The liability accrued in a past accounting period for a product recall (to correct a defect) now appears to be inadequate and must be increased (and the corresponding expense recorded in this period).

- The taxing authorities are conducting an audit of the company's recent tax returns and the company anticipates that this audit will result in the assessment of additional taxes.

AGAIN, FEDERATED DEPARTMENT STORES

Return to Federated Department Store's balance sheet, Exhibit 1-2 in Chapter 1. Timing issues arise with respect to many of the assets and liabilities listed. The word "net" that completes the label for amounts due from customers, accounts receivable, signals that Federated carries an allowance for doubtful accounts; the notes to the financial statements reveal its amount, $113 million, or 3.2% of gross accounts receivable. Clearly, Federated has experienced in the past, and therefore anticipates in the future, bad debt losses equal to about 3% of credit sales. The word "net" also completes the labels for property & equipment and for other intangible assets. Once again, the financial notes reveal the amounts of the accumulated depreciation and amortization, respectively; the depreciation and amortization methodologies, the useful lives, and the original costs of these long-term assets are also spelled out, as well as the resulting depreciation and amortization expenses appearing in the 2004 income statement.

In the current liabilities section, accounts payable and accrued liabilities are combined. The notes tell us that amounts owed to vendors, accounts payable, were $1,301 million at year-end, and thus accrued liabilities were $1,406 million. Accrued liabilities generally reflect employee-related expenses incurred but not yet paid: for example, unpaid salaries and wages; income taxes withheld on behalf of employees but not yet remitted; fringe benefits such as medical insurance, disability and workmen's compensation insurance, vacations earned but not yet taken, and perhaps performances bonuses earned in fiscal year 2004 but not yet paid. Employee salaries and wages undoubtedly dominate the income statement line item "Selling, General and Administrative" expenses; the accrued liabilities at year-end are equivalent to about 28% of the total of those expenses.

Jumping the gun a bit in Chapter 7, note that the $1,301 million of accounts payable equals a bit less than 15% of the $9,297 cost of sales incurred last year (see Exhibit 2-2). We might conclude that Federated owes about (0.15 × 52 weeks) 7 weeks worth of purchases, say 50 days. Is Federated paying its vendors in a timely manner? Do not forget that the department store business is quite seasonal and the $1,301 million accounts payable balance is as of January 29, right at the end of its busiest season.

CHALLENGING TIMING ISSUES IN CERTAIN INDUSTRIES

Certain types of businesses present some particular challenges regarding the timing of revenues or expenses. The explanation of a few of these should help illustrate the importance of timing considerations.

Book Publishing

Most expenditures associated with publishing a new book are incurred before the first volume is sold: acquiring the manuscript, editing, typesetting, printing the first run. Should these expenditures be treated as expenses in the periods when they are made? Or, alternatively, should they be matched against the revenue generated from the sale of the book? The principle of matching suggests the latter. Accordingly, a major portion or all of these expenditures must be *capitalized* (debit to an asset and not to an expense account when they are incurred). Now step forward to the accounting period when revenues from the book's sale are first received. How much of this asset should be matched against that revenue? Ideally the cost of goods sold per book would be the same throughout the life of the book. But that ideal requires that the publisher estimate accurately total book sales, and do so just as sales commence and only sketchy sales forecast data are available. Inevitably, book publishers will sometimes overestimate and other times underestimate sales. As additional sales data are available, forecasts are refined and appropriate adjustments are made to accounting for future cost of goods sold. With experience, publishers' forecasts become quite accurate, but that does not mean they will not periodically be surprised—in both directions! And, obviously, the principle of *conservatism* suggests they should lean in the direction of underforecasting.

Motion Picture Studios

Motion picture studios have a similar challenge, but more severe than book publishers. The studio produces relatively few movies per year; each tends to cost very large sums; movie critics stir up "buzz" that can turn the picture into a blockbuster or a dud; much of the promotional costs as well as the production costs are incurred before the movie is released; revenues come in several forms, including the sales of DVDs well into the future; and so forth. Unsurprisingly, earnings at motion picture studios tend to be very volatile from accounting period to accounting period.

Software

These same conditions obtain in the software industry, but here some specific accounting rules have developed which require that the producer of software declare a moment in time when "research" has been completed (the phrase is when "technological feasibility has been demonstrated"). Expenditures up to that time are expensed; expenditures after that moment are capitalized and matched against future revenues from the sale of the software. Obviously, management has broad latitude in determining when "expensing" should cease and "capitalizing" should commence.

Industries Offering Generous Product Warranties

Suppose that a particular manufacturer is so confident of its product that it offers full refunds during the first 6 months to any customers not satisfied with the product. This offer amounts to an exceedingly generous warranty. The manufacturer is not warranting specific performance, but simply "satisfaction." Question: when should the manufacturer recognize the revenue from the sale, immediately or only at the end of the 6-month money-back-guarantee period? The conservative choice is obvious. But suppose that this manufacturer turns out to have extraordinary success with the product, customers are very satisfied, and few or none have requested a refund. Under these conditions, would not a 6-month postponement in the recognition of revenue amount to excessive conservatism?

MORE ACCOUNTING RULES

In summary, let us reemphasize a couple of the valuation rules outlined in Chapter 3 and augment the list with the rules introduced in this chapter relating specifically to timing issues.

- *Conservatism* is as powerful a principle in determining the timing of accounting entries as it is in determining their value. To repeat, conservatism requires that accountants lean in the direction of early reporting of expenses and delayed reporting of revenues.

- Implementing the principle of *matching* causes the *deferrals* and *accruals* discussed in this chapter.

- The *consistency* principle refers to consistency *within* a corporation, not between corporations, and requires that accountants not change their minds willy-nilly about accounting policies. For example, once a company decides to account for cash discounts on the net basis or use accelerated depreciation in valuing fixed assets, it needs to stick with that policy. This principle enhances comparability of financial statements across accounting periods, thus facilitating ascertainment of trends in financial performance or position.

- The *materiality* principle saves accounting work without sacrificing the usefulness of the financial information.

NEW TERMS

Accrued expense or revenue

Accelerated depreciation

Adjusting entries

Allowance for depreciation

Allowance for doubtful accounts

Amortization

Book value (of fixed assets)

Cash discounts, gross/net accounting

Contra accounts

Consistency

Declining balance depreciation

Deferred revenue or expense

Depreciation

FIFO (first-in, first-out)

LIFO (last-in, first-out)

Materiality

Straight-line depreciation

Warranty

CHAPTER 5

Capital Structure
Using Debt Prudently

Our balance sheet focus thus far has been dominantly on current assets, fixed assets, and current liabilities. We have said little about how corporations are financed or capitalized. This chapter focuses on the permanent capital used by corporations: capital obtained by selling stock to shareholders, by retained earnings, and by borrowing on a long-term basis. That is, our attention turns to the long-term debt and owners' equity sections of the balance sheet.

A thorough discussion of financing methods can be found in any of a host of corporate finance textbooks or trade books. Here we can only introduce some key topics, taking the perspective of the corporation rather than that of investors.

Long-term debt and shareholders' equity are considered the corporation's permanent sources of funding. Current liabilities also supply funds (for example, short-term bank borrowing and accounts payable), but these sources tend to expand and contract in proportion to the volume of business activity. Current liabilities would dry up if the corporation ceases doing business; of course, so also would the current assets; thus, they are not considered permanent.

SOURCES OF DEBT

Money can be borrowed from many sources, including the owners' relatives, commercial banks, trade creditors, insurance companies, private investors, and the public market. I will not attempt to generalize about borrowing from your relatives; amounts you could borrow and terms and conditions depend on who your relatives are and what they think of you and your business! Let us look at the other debt sources.

Trade Credit

Terms of sale in a large majority of commercial transactions are "net 30 days." The buyer is expected to pay in 30 days from the date of shipment of the goods or performance of the services, but of course many buyers do not. A few pay more quickly and others stretch their payments to 45, 60, or 90 days. If Corporation L shifts from paying in 30 days to paying in

60 days, its accounts payable will double with the same rate of purchasing activity. These are useful funds for Corporation L, all the more so because L's vendors typically will not charge interest, as the bank would (but L may forego cash discounts for prompt payment). In effect, this is free borrowing.

But is it really "free"? Corporation L's vendors may be unhappy with this new pattern of slower payments. In addition to hounding L to speed up payments, these vendors might take other subtle actions against L, since L is now not a favored customer: offer less advantageous pricing, reduce L's shipping priority, withhold or delay information on new or enhanced products. If a vendor gets sufficiently annoyed at L's slow-paying habits, the vendor could decide to drop L as a customer, or service L only on a COD (cash on delivery) basis–that is, cease extending credit to L. When one vendor takes such action, word quickly spreads in the industry, and other vendors follow suit—and L soon finds that its "free" source of borrowing (accounts payable) dries up, often with dire results for L.

Other Short-Term Borrowing

Historically, short-term bank borrowing was used only to meet seasonal needs; for example, retailers borrowed in the months leading up to their busy seasons, pumpkin growers paid off their short-term borrowing after Halloween, and building contractors borrowed in the spring. These loans were considered to be "self-liquidating"; at the end of these busy seasons, inventories and accounts receivable balances would return to lower levels, freeing up cash that would be used to repay the short-term borrowing. In recent decades, banks have become more aggressive—or liberal, if you like—and no longer require that short-term loans be "cleaned up" annually.

Very creditworthy companies can borrow short term on an unsecured basis; they need provide no collateral to the banks. Any bank would love to lend to Microsoft, General Electric, Federated Department Stores, and similar companies with very strong financial positions as revealed by their balance sheets.

Less creditworthy corporations may need to provide **collateral** for any borrowing, including short-term borrowing. What is the best collateral that it can offer? Cash, of course, but if it had a lot of cash, it would not need to borrow. The next best candidate is accounts receivable, the current asset that will very soon turn into cash. Often banks will lend up to 80% of the accounts receivable balance, generally deducting from that balance any receivables that are, say, 90 days past due.

The next most liquid current asset is typically inventory, and its value as collateral is entirely a function of the nature of the inventory. An auto dealer readily borrows against its inventory of new cars; a metals wholesaler can borrow against its inventory of standard steel and aluminum shapes; and retailers can borrow some percentage of their inventories of standard, readily saleable merchandise, but perhaps not fashion or "fad" merchandise. The more specialized the inventory, the less value it has as collateral. Consider, for example, a manufacturer where inventories are dominantly "in process," neither raw materials nor finished goods. In-process inventory is useless as collateral since the lender can realize little or nothing from its disposition, should the lender have to repossess it.

Companies with poor credit ratings may still be able to borrow short term, but on rather unfavorable terms. The lender will not only charge a high interest rate but also place other restrictions on the borrower such as limiting the purchase of fixed assets, prohibiting other borrowing or leasing, forbidding the payment of dividends, and placing a ceiling on executive pay. Specialized lenders not only charge high interest rates and fees but insist that they have facilitated access to the collateral: factoring companies buy receivables at a discount, insisting that customer payments come to the factoring company directly; still other lenders maintain physical control over collateralized inventory.

Federated Department Stores borrows short term a total of $1,242 million from a group of commercial banks. Notes to its financial statements reveal that the borrowing is secured by the company's accounts receivable. Other short-term borrowing agreements permit Federated to borrow short term up to a total of $2.4 billion on an unsecured basis. These agreements represent a cash cushion for Federated; the company was borrowing nothing under these agreements as of January 29, 2005. The banks have placed minimal other restrictions on Federated; they obviously consider Federated to be an excellent customer!

Long-Term Borrowing

Long-term loans, by definition, involve principal repayments—part of or the entire principal—a year or more in the future. For example, a 5-year term loan may provide for equal annual repayments of principal over the 5 years. Such a loan might be collateralized, perhaps by fixed assets, or unsecured if the borrower has a strong balance sheet. Generally, term loans have restrictions or "covenants" that permit the lender to "call" the loan—demand immediate repayment—if certain milestones are not met or other unfavorable events occur. Interest rates can be either "fixed"—unchanging for the entire term of the loan—or "floating," tied to a widely published rate (e.g., bank prime lending rate) that fluctuates as credit conditions change.

It is worth reminding ourselves that the demand by a lender that the borrower immediately repay a short- or long-term loan does not assure that the lender will receive its funds. Seldom does a borrower have sufficient cash on hand to effect immediate repayment, and only quite creditworthy customers can turn to another lending source to replace the original lender. Very often lenders and borrowers will have to cooperate in "working out" the troublesome loan. The lender is not eager to force the borrower into bankruptcy even if, as is generally the case, the lender has that option. The lender is likely to suffer greater loss as a result of bankruptcy proceedings than from a "work out."

Generally, a borrower pays higher interest rates on long-term loans than on short-term borrowings. The notes to Federated's financial statements reveal that Federated's long-term debt consists of eight different issues, all unsecured, with interest rates ranging from 6.6 to 8.5% and due at various dates from 2008 to 2029. Federated's short-term debt is secured by its accounts receivable. The average interest rate paid by Federated in 2004 on all of its borrowing was 7.7%.

Leasing

Another source of long-term funds—closely equivalent to borrowing—is **leasing**, typically of fixed assets: buildings, machinery, vehicles, and instruments. Title to the fixed asset remains with the leasing company (**lessor**), and the user (**lessee**) makes monthly lease payments. Some leases are really rentals—for example, many copying machines are rented; the lessee can terminate the lease on short notice. On the other hand, leases with terms about equal to the useful life of the fixed assets are referred to as **capital leases** and are generally equivalent to installment purchases: the lessee does not have the option to terminate the lease and over the full term of the lease the lessee will, in effect, pay the purchase price of the fixed asset plus interest. At the end of the lease, the lessee can typically purchase the asset for a residual value that is often minimal.

Why would a corporation lease a fixed asset rather than purchase it? Companies with weak balance sheets often find that they can lease when they are unable to borrow the equivalent amount. Compared to the position of a lender, the leasing company has the advantage of retaining legal ownership of the fixed asset; repossession, if necessary, is simpler. Historically, companies could dress up their balance sheets by not recording long-term leases: showing neither the value of the fixed asset nor its obligation to the leasing company. More recently, accounting rules have been revised so that both the asset and the obligation, typically valued on a time-adjusted basis (see Chapter 3), must be recorded if the lease is judged to be economically equivalent to a purchase/borrowing transaction. Thus, that particular advantage of leasing has disappeared. Finally, if a company is unable to utilize the depreciation tax shield—for example, a loss-making airline or a start-up company that will incur losses in its early years—it can often gain an advantageous leasing rate from a leasing company that can utilize the tax shield.

Public Borrowing

Large companies borrowing large sums also have access to the public debt market. That is, they issue and sell to a variety of investors their promise to pay, typically referred to as a

bond or **debenture**. Such debt issuances are typically **underwritten** by investment banks that undertake the distribution of the bonds; buying the bonds from the issuing corporation at a small discount, the underwriters then assume the risk that they can be sold at the agreed-upon price, often to a great number and variety of investors. After issuance, the bonds trade in organized markets; their prices fluctuate to reflect both changes in prevailing interest rates and any improvement or deterioration in the credit soundness of the issuer. The primary purchasers of such bonds are institutions: mutual funds, insurance companies, pension and endowment funds, and banks. Individuals, however, are certainly not precluded from purchasing corporate bonds.

Publicly issued bonds tend to have reasonably long maturities, say 10–30 years. Some require periodic repayment of portions of the principal but some call for a "balloon payment," the full amount (or a large percentage thereof) due on the maturity date. Some bond issues are secured, perhaps by land and buildings or by customer receivables; others are unsecured. Bonds can be denominated in Euros, Yen, and certain other international currencies, as well as in U.S. dollars.

High-Yield (Junk) Bonds

Starting several decades ago, a new market for bonds emerged, the so-called **junk bond** market, now more genteelly called the **high-yield** market; junk bonds are contrasted with investment-grade bonds. Companies with quite weak creditworthiness can often borrow in the high-yield market by offering substantially higher interest rates (coupon rates) and tighter restrictions. For example, when investment-grade bonds offer coupons of 7%, high-yield bonds may carry 12–15% coupons. Investors willing to accept higher risks to earn high returns are attracted to the high-yield market.

Recognize that investment-grade bonds can turn into junk bonds. If, following issuance, the issuing corporation falls on hard times and its credit rating deteriorates, the price at which its bonds trade in the market will fall, with a corresponding increase in the bonds' yield (i.e., effective rate of interest; see Chapter 3) until the yield is in the range of other high-yield bonds. Conversely, high-yield bonds can morph into investment-grade bonds if and when the issuer's financial health improves; then, the bonds' market price will appreciate, sometimes quite markedly. This opportunity for price appreciation is a major attraction for high-yield bond investors.

Remember that changes in market price do not alter the valuation of the bonds on the issuing corporation's balance sheet. However, some corporations may find it advantageous to buy up part or all of its publicly traded bonds if an increase in prevailing interest rates has driven the bonds' market prices down significantly.

EQUITY

As you know, corporations are owned by their common stock shareholders who elect the board of directors that, in turn, hires, guides, oversees, decides the compensation of, and, as required, replaces the corporate officers. In the United States the organizers of a new corporation apply to a state, not necessarily the state in which its headquarters will be located, for permission to operate. Incorporating is reasonably straightforward and inexpensive. The founders may contribute assets other than cash—intellectual property in the form of patents or know-how, or fixed assets, or the assets of a predecessor partnership that is now being incorporated—in return for shares of common stock. All corporations have issued and outstanding common stock.

Other forms of business organization are sole proprietorships and partnerships; these are in effect extensions of the sole proprietors or partners themselves and require no governmental permission to commence business. Since corporations conduct the vast majority of business, we will concentrate our attention on that form.

Corporations may from time to time (and not solely at the time of incorporation) sell newly issued common stock to raise funds for the business. When the intended investors are the general investing public, the sale is typically underwritten by a syndicate (group) of investment banks. Underwritings are highly regulated by government agencies (see Chapter 10). Investors buy these shares seeking returns in one or both of two forms: dividends and appreciation in the price of the shares. Over the great sweep of the last 100 years or so, common stock investments have returned a combined total of about 9–10% per year—somewhat less than many investors assume—but annual returns have varied widely around this average. Dividends are not assured, but are paid at the sole discretion of the board of directors; many companies as a matter of policy do not pay dividends, instead investing all available cash in growing the company (often including buying other companies). At dividend-paying corporations, directors "declare" dividends, most typically each calendar quarter, at so-many cents per share. Of the 500 largest companies in this country, 385 paid dividends as of early 2006; this number is down substantially from 25 years ago, but up somewhat over the last 5 years.

These characteristics of common stock define why the sale of common shares is for the corporation much less risky than taking on an equivalent amount of additional debt. It is also true that capital raised by selling common stock is typically more expensive than borrowed capital.

Most corporations start their lives with private placements of equity. The investors may be friends or relatives, but some financial institutions (typically called venture capital firms) exist for the purpose of making early-stage investments in promising new companies. Venture capitalists invest in high-risk situations offering the prospect (but certainly not the promise) of very high returns, say 25% or more per year. The venture capital market in this country, as compared with other countries, has become very large, increasingly well segmented, and is the key to our nation's robust entrepreneurial activity. Individual venture capital firms may focus on

- start-up companies or more mature companies needing additional funding;
- small investments (often called seed-capital investments) or large investments;
- technology companies or service companies or internet companies or biotech companies;
- certain geographic regions.

In addition to supplying capital, the best venture capitalists provide young firms with significant assistance in operations, in hiring for key positions, and in acquiring additional capital as the company grows.

In recent years so-called private equity firms have emerged with huge resources to invest in mature companies, acquiring major positions up to 100% of the shares. Typically, these firms will restructure the acquired operations (sometimes these are divisions of conglomerate firms) and after a few years seek to free up their investments through one of several so-called exit strategies.

STOCK MARKETS

In this country, it is hard to escape some familiarity with the stock markets. The media report daily—often breathlessly—on price movements in these markets, the major ones being the New York Stock Exchange (NYSE) and NASDAQ, sometimes referred to as the over-the-counter market. All developed economies and many developing ones also have active stock markets where securities are bought and sold. And, of course, a huge industry of brokers, financial advisors, underwriters, and analysts has developed to assist and facilitate investors (and speculators) in these stock markets.

Let me emphasize again that trades executed on these markets are between shareholders and thus have no direct effect on the financial statements of the companies whose shares are being traded. A company's executives and directors, of course, are not indifferent to the price movements of their shares even if they have no definitive plans to issue and sell new shares. Shareholders who have lost money on their investment are frequent agitators for changes in management or the board.

Another definition to remember is that of **market capitalization** (often referred to as simply "market cap" of the corporation): the product of the market price per share times the number of shares outstanding. This is the value that the stock market assigns to the total ownership of the corporation. Remember that market capitalization bears no necessary relationship to the total owners' equity value shown on the balance sheet (that is the **book value** of equity). For example, Federated's market capitalization has been running somewhat above its book value.

When is a Stock a Good Candidate for Purchase or Sale?

We will return to this question in Chapter 7 but I need to disabuse you now of any notion that some particular piece or pieces of financial information available in the company's published financial statements will tell you when to buy and when to sell.

Stock prices go up when demand for shares by potential buyers exceeds the supply of shares from potential sellers—hardly a profound statement but a useful one to bear in mind. Shares of a particular company are a good buy for you if your expectations about the future of the company are rosier—you are more bullish about future profits and dividends—than the expectations held by the investor universe in general; and vice versa. If your expectations are born out over the next months and years, the stock price will go up as other investors catch on; you will benefit from that price appreciation. Financial statements help you form your expectations about the company's future.

STOCK DIVIDENDS AND STOCK SPLITS

Boards of directors may from time to time decide to **split** the company's **stock** or declare a **stock dividend**. These decisions lead to a lot of paper shuffling but are not of great significance to an individual shareholder. If Company D's stock sells for $80 per share just before D's directors declare a four-for-one stock split, it will sell for $20 after the split. Each shareholder will have four times the number of shares previously owned, each worth one-fourth as much. Similarly if D's board declares a 10% stock dividend, each shareholder will receive one additional share for each 10 shares now owned, and we can anticipate that the price of the stock will decline about 10%. Thus, if the market acts totally rationally (a somewhat naive notion!), each shareholder owns the same percentage of the company before and after the stock split or dividend and the aggregate market value of his or her investment is unchanged.

But there are two reasons why companies declare stock dividends or splits. One is a "signaling" reason: the declaration itself sends a signal to the investing world that the directors are optimistic about the future of their company and its share price. The second is that a stock split can be used to reduce the market price per share into a range that is deemed more attractive for investors; a conventional wisdom is that shares selling in the range of $20–50 per share draw more investor attention than those trading outside that range. (Incidentally, "reverse splits" are used to increase the per-share stock price for companies whose shares are trading at what the directors believe is too low a price, say less than $2 per share.) Note too that if a company holds its dividends per share steady through a 10% stock dividend, it has in effect increased dividends by 10%.

STOCK OPTIONS AND WARRANTS

Stock options are widely used as an incentive for key company employees. These options permit (but do not require) the holder—that is, the employee—to purchase the allotted number of shares at a set price for up to, say, 5 years into the future. That price is typically the market price on the day the option is granted. An option is an attractive one-way street for the option holder: he or she has no money at risk until the option is exercised and the option will be exercised only if or when the share market price is comfortably above the option price. The idea is that stock options should align the motivations and incentives of the employee-optionee with those of the shareholders.

On January 29, 2005 Federated had outstanding employee stock options totaling just under 20 million shares, equal to more than 10% of total outstanding shares. Of these, about half were exercisable (i.e., restrictions no longer applied) and at an average exercise price above $40 per share.

We will see in Chapter 10 that controversy brewed for years as to how the "cost" of these employee options should be reflected in financial statements. Some take the rather extreme view that employee options are "costless."

A **warrant** is simply an option granted to a nonemployee. For example, a bundle of warrants might be granted to a lender in connection with a major, long-term loan to the company.

EARNINGS PER SHARE (EPS) AND PRICE/EARNINGS (P/E) RATIO

If Company P's shares today have a market price of $30 and Company Q's market price per share is $20, is Company P's stock "more expensive"? We have just said that market prices can be manipulated—adjusted may be a more appropriate word—by stock splits and dividends. You can make the same investment in these two companies by purchasing 200 shares of Company P and 300 shares of Company Q. We need some other measure of "expensiveness" and that is the **price/earnings (P/E) ratio**.

The measure of profitability most relevant to a shareholder (present or potential) is not total net income but rather net income per share (or **earnings per share**, abbreviated **EPS**). EPS is simply net income divided by the number of common shares outstanding. That value divided into the share price results in the price/earnings ratio for the stock.

Now we can say that if Company P's price/earnings ratio is 15 and Company Q's is 20, Company Q's stock is more expensive even though its market price per share is only two-thirds that of Company P. Think of it this way: an investor buying Company Q's stock is willing to pay 20 times the current annual earnings attributable to each share, while the investor in Company P is willing to pay only 15 times. This investor must have greater expectations for the future of Company Q than of Company P.

Federated Department Stores shows as the last item on its published income statement for the year ended January 29, 2005 earnings per share for the year of $3.93. Notes to its financial statements indicate that 167 million shares of Federated common stock are outstanding at year-end. Federated's net income of $689 million divided by the average number of outstanding shares during the year yields the $3.93 EPS figure.

Federated's published income statement also includes a value for **diluted earnings per share**, $3.86. This lower EPS gives effect, in a complicated calculation, to the exercisable but unexercised stock options that will, when exercised, increase the denominator of the EPS calculation and thereby reduce EPS—the equivalent of about 3 million additional shares. The difference between basic and diluted EPS increases as (a) the number of exercisable stock options grows and (b) the difference between current market price per share and average exercise price per share increases. Companies that have experienced large increases in share price and use stock options as a major form of compensation will end up with large differences between basic and diluted EPS.

GOING PUBLIC

Most small companies are privately owned; that is, their shares do not trade on either of the markets mentioned earlier. Of course, shares can be sold between individuals in privately negotiated transactions (subject to certain legal restrictions). At some point the company may decide to "go public," that is, sell shares to a broad set of investors and thereafter comply with the legal requirements to permit public trading of the shares following the **initial public offering (IPO).**

Why would a company decide to go public? Two primary reasons: to raise substantial capital from a new set of investors—an amount that existing and other private investors are unwilling or unable to provide—and to provide liquidity for its present shareholders (who may include the company's founders and venture capitalists). By liquidity we mean the opportunity to sell shares in an orderly market; this is a popular exit strategy for venture capitalists who are eager to exit Company A when it has reached a certain degree of maturity and invest those proceeds in another small high-risk/high-potential-return enterprise. And founders are often eager to sell some of their shares so as to better diversify their personal investment portfolios.

SHARE REPURCHASE

Remember that a corporation can, upon the decision of its board and with appropriate public notification, go into the market and repurchase its own shares. In recent years a great number of companies whose shares trade publicly have bought back huge numbers of shares. This is a rational investment decision if the board feels that its current market price is unrealistically low. And, the purchases themselves add to the demand side of the equation; if supply is unaffected, then elementary economics tells us that the market price per share should rise. Note also that the company's EPS, assuming no change in net income, will increase as shares are repurchased. And, if the P/E ratio is unaffected by the repurchase, the market price per share will increase as EPS increases.

THE ROLE OF REGULATORS

All that we have been discussing—share trading, IPOs, stock repurchases—are subject to comprehensive and complicated regulation, primarily by the Federal government (particularly the Securities and Exchange Commission, the SEC) and to some extent by state governmental agencies. Note that the SEC's name signals that it regulates both securities and the exchanges on which they are traded. A full explanation of these regulations would require an entire book, but suffice to summarize here some of the major purposes of these regulations:

- To insist on full and accurate disclosure. The SEC does not try to shelter investors from making foolish (even stupid) decisions, but only to assure that investors are provided with full disclosure of all facts deemed relevant to their investment decisions. For example, in IPOs the SEC does not opine on the quality of the company or the investment opportunity being offered. The resulting "disclosure documents" are often so long and so full of legalese that few investors, unfortunately, actually avail themselves of the full benefits of this comprehensive disclosure.

- To minimize conflicts of interest, and disclose those that cannot be eliminated. Share trading is replete with conflicts of interest between investors on the one hand and, on the other hand, brokers, exchanges, investment counselors and others offering services to those investors.

- To assure that unsophisticated investors—historically and unkindly characterized as widows and orphans—are protected from unscrupulous companies, brokers, and others in the financial industries.

- To minimize fraud by exacting stiff penalties when fraud is uncovered.

- To support the establishment of sound accounting policies and promote their use by all companies whose securities are traded publicly.

- To insist on sound financial controls sufficient to assure accurate financial reporting.

PREFERRED STOCK

Thus far in this chapter we have discussed debt and common stock securities. Another class of securities with some of the characteristics of each is **preferred stock**. Like debt securities, preferred shares carry a specified return—called a dividend rather than interest. Unlike interest on debt, however, a company's failure to pay preferred dividends, while serious, is not catastrophic. Typically, when preferred dividends are not paid on schedule they "cumulate": they are a continuing obligation of the issuing corporation and must be paid before dividends on common stock may be paid. And, unlike common stock, preferred shares do not carry the privilege of voting for the board of directors; except, when unpaid preferred dividends have accumulated, the preferred shareholders are often given voting power.

If common stock is a less risky but more expensive source of new capital as compared with debt, it follows that preferred stock is in the middle: more risky and less expensive than common stock, and less risky but more expensive than debt.

HYBRID SECURITIES

The discussion above by no means exhausts the forms of security that can be used to raise capital for corporations. Investment bankers are constantly inventing new hybrid securities, having some characteristics of debt and other characteristics of common stock. A widely used hybrid is a security that is nominally either debt or preferred stock but which is **convertible** into shares of common stock at certain times and under certain conditions. A quick example is as follows. Suppose Company Z needs additional capital and is debating whether to sell new debentures (debt) or issue and sell more common shares. Z's directors feel that the common share price at $25 per share is unattractively low and the anticipated interest rate on debentures at 8% is unattractively high. Z might decide to sell convertible bonds—each $100 bond being convertible into, say, three shares of common stock. Because of this "equity kicker"— that is, the opportunity for the buyers of the bond to become common shareholders if Z prospers—Z is able to sell the bonds with a 6% interest (coupon) rate. If the market price of Z's common shares appreciates beyond $33.33 per share, the conversion privilege is attractive—the conversion is "in the money"—and Z will, in effect, have sold the new shares for $33.33 rather than $25.

A final caveat in our discussion of forms of securities (debt, common equity, and preferred stock): each has unique income tax implications for investors that influence their relative attractiveness. In short, "see your tax accountant!"

WHEN IS BORROWING MORE ATTRACTIVE THAN SELLING ADDITIONAL EQUITY?

If debt funds are cheaper than equity funds, why do companies not use more debt? If funds obtained through the sale of common stock are less risky, why do companies ever use debt?

Those two questions answer each other. Obviously, some combination of debt capital and equity capital appears to be ideal: balance risk and cost. Before we consider how to make that tradeoff, look first at the advantage of using debt.

DEBT LEVERAGE

Debt leverages—that is, it can potentially push up or push down—returns to the common shareholders. We say that a company that uses lots of debt is highly **debt leveraged**. A simple example: if Company D can borrow long term at an interest rate of 7% and can invest those funds in projects that earn an 11% return, the incremental 4% accrues to the benefit of the shareholders.

Compare the following two companies with quite different capital structures (proportions of debt and common equity) but identical operating results, as measured by earnings before interest and taxes:

	Company B	Company C
Debt (7% interest rate)	$400,000	$800,000
Equity	900,000	500,000
Total capital	$1,300,000	$1,300,000
Earnings before interest and tax	$200,000	$200,000
Interest expense	28,000	56,000
Income before tax	172,000	144,000
Income tax (40% rate)	68,800	57,600
Net income	$103,200	$86,400
Net income as percent of equity	11.5%	17.3%

Company B generates more dollars of net income, but that is not what counts to shareholders. Company C earns a higher return on equity than Company B; with half the number of shares outstanding, C's EPS is higher than B's. In short, Company C's shareholders' return has been positively leveraged by its greater use of debt.

Now suppose that both companies encounter a recession and their earnings before interest and tax fall to $80,000 while their total capital remains as shown. Note that $80,000 is 6.2% of total capital; that is not enough to pay the 7% interest on the debt portion of that total capital. Thus, both companies' shareholders will suffer negative leverage, but C's will suffer more. In fact, under this scenario, B's net income is higher than C's and, more importantly, B's net income as a percent of equity is 3.5%, slightly higher than C's return of 2.9%.

And, of course, if the recession turns very severe and both companies' net income turns negative, the interest on debt still must be paid. Company C, with the greater interest expense, will get in financial trouble sooner than will Company B. We say that Company C, with higher debt leverage, has a riskier capital structure.

EQUITY DILUTION

When a corporation sells additional common stock, it suffers **equity dilution**. That is, existing shareholders each own a slightly lower percentage of the total company; their ownership position has been diluted. The number of outstanding shares—the denominator of the EPS fraction—increases, thus lowering EPS. Of course, if the cash raised from the sale of new shares is invested in projects that boost net income, the EPS dilution may be short lived.

In addition to EPS dilution, issuing additional common shares also dilutes ownership control by increasing the number of shares required to exercise voting control. If a company's directors and executives own small fractions of the outstanding shares, they may view this ownership dilution as a positive event, since major shareholders will now have a tougher time gathering enough votes to remove directors and take over the company. On the other hand, if the board members and executives together own more than 50% of the outstanding shares, they may be very reluctant to undertake any financing that would seriously dilute their ownership, particularly below that threshold. (Incidentally, at large corporations where stock ownership is widely disbursed, it is possible for a shareholder or a group of shareholders owning substantially less than 50% of the shares still to exercise effective control.)

DESIGNING A CORPORATE CAPITAL STRUCTURE

This discussion of debt leverage and dilution suggests that for any company there ought to be an optimal capital structure: a combination of debt and equity that properly balances risk and return. The primary factors to consider in designing a company's capital structure are as follows:

- How steady are the company's earnings over the course of a business cycle? An electric power utility, with quite steady returns, can and should utilize high debt leverage to improve its returns to shareholders. In contrast, a manufacturer of heavy construction equipment can expect its revenues and earnings to fluctuate widely over the course of a business cycle; it should minimize the use of debt leverage.

- How risky is the business? A high-technology company faces considerable technology risk, the risk that some or all of its products become obsolete overnight due to a competitor's technological breakthrough. The high-tech company will probably want to avoid compounding its risks, that is, adding the risk of high debt to the unavoidable high-technology risk. On the other hand, a consumer products company with strong brand loyalty may have low operating risk and therefore can afford to borrow a higher percentage of its total capital.

- How risk tolerant are the corporate officers, the directors, and the shareholders? The question is whether these folks choose to sleep well or eat well. Some people just do not sleep well when they are exposed to high risk. Others are risk tolerant and give higher priority to improving shareholder returns. The first group will avoid borrowing; the second group may be comfortable borrowing as high a percentage of its total capital as possible, limited only by lenders' ultimate refusal to lend more.

- Are ownership control issues important? A concern about voting control pushes a corporation toward the greater use of debt capital and the avoidance of new equity financing.

CORPORATE MERGERS AND ACQUISITIONS

Many corporate mergers and acquisitions—often referred to as M&A activity and more formally and generically as business combinations—occur daily. A merger involves a combination of two companies of similar size into a single combined corporate entity; shareholders of both companies receive shares of the newly created company. Acquisitions are paid for with (a) cash, (b) debt of the acquiring company, (c) common stock of the acquiring company, or (d) some combination of (a), (b), and (c). Some are "friendly"— the acquiring company and the acquired company negotiate a price and cooperate in effecting the corporate combination; and some are "hostile"—the acquirer seeks to take over a reluctant target company. Hostile takeovers are generally accomplished in two steps: the acquirer buys in the open market a significant number of shares in the target company and then makes a "tender offer" to acquire the remaining outstanding shares, all without the consent of the management and board of the target. Unsurprisingly, top executives of companies taken over this way rarely keep their jobs.

Many studies have shown that the majority of corporate combinations—mergers or acquisitions—are unsuccessful. That is, the shareholders of the surviving company are worse off after the merger than they were before. Nonetheless, M&A activity continues at a fast pace, perhaps because more attention is paid to simply growing than to improving shareholders' wealth (though the latter is the corporation's legal obligation). As some wags say of marriage, an acquisition or merger is a triumph of hope over reason.

How do we account for acquisitions? The acquiring company is assumed to be buying the assets of the acquired company; they are valued at fair market value on the acquiring company's balance sheet. Typically, the acquiring company assumes most or all of the liabilities of the acquired company; this assumption of liability is, in effect, an addition to the purchase price paid. The revalued assets (including identifiable intangible assets) and liabilities are taken on to the acquiring company's books. If the purchase price exceeds the net of the assets obtained and the liabilities assumed, the difference is assigned to the asset Goodwill, discussed briefly in

Chapter 1. Large companies that make a steady stream of acquisitions often accumulate very large Goodwill balances.

Is goodwill then amortized as other intangible assets typically are? The answer is: maybe. Rules about goodwill amortization have changed frequently—from no amortization to amortization over a fixed number of years. Remember that amortization expense serves to reduce net income (but not cash) for the period; accordingly, companies are not eager to record large goodwill amortization expenses. The current rule requires that the company reevaluate its goodwill each time it publishes a balance sheet; if the goodwill has become "impaired," that impairment should be reflected by a reduction in the value of goodwill. The rub is: how does one determine impairment? If an acquired operation that resulted in recording substantial goodwill is later terminated completely, one can safely assume that the associated goodwill is so impaired that it has become worthless. But if that same operation simply deteriorates slowly over many years, assessing year-by-year impairment of goodwill is difficult if not impossible.

ONCE AGAIN: FEDERATED DEPARTMENT STORES

The notes to Federated's January 29, 2005 financial statements reveal that soon thereafter—in February 2005—Federated entered into an agreement to acquire. The May Department Stores Company for which it expects to pay $5.5 billion in cash plus 95.9 million shares of Federated common stock. Here a company with about $15 billion in total assets and $16 billion in sales is acquiring a company that last year had $15 billion in assets—about the same as Federated— $14.4 billion in sales and net income equal to 76% of Federated's net income. This is a very major acquisition for Federated.

As of January 29 Federated's cash balance was $868 million, well short of the $5.5 billion needed for this acquisition. Presumably, Federated is confident that it can raise the needed additional cash through new financing. Will Federated be able to borrow an additional $5 billion or more, pushing its long-term debt to $7.5 or $8 billion? Is additional debt leverage desirable for Federated, or might its financing risk rise to unacceptable levels? These are surely questions that Federated's board of directors thought hard about before agreeing to acquire May.

NEW TERMS

Book value (of owners' equity)	Debenture
Bond	Debt leverage
Capital lease	Diluted earnings per share
Convertible security	Equity dilution

Lease

Lessee

Lessor

Earnings per share (EPS)

Market capitalization

Initial public offering (IPO)

High-yield bonds

Junk bonds

Preferred stock

Price/earnings ratio (P/E)

Stock dividend

Stock option

Stock split

Underwriter (underwritten)

Warrant

CHAPTER 6

Cash Flow
Cash Is King!

Chapters 1 through 5 focused on the balance sheet and the income statement—what we referred to as the fundamental documents of the accounting system. Early on we acknowledged that cash inflows and outflows very frequently do not occur at the time when we recognize, respectively, sales/revenues and expenses. We quickly concluded that we needed to use accrual accounting to do the proper job of matching—matching cost of goods sold to sales, and matching revenues and expenses to the relevant accounting period—and thus to arrive at a proper valuation of net income, assets, and liabilities.

But net income cannot be used to pay bills, salaries, principal on loans, dividends, and so forth. Cash is required for those purposes. It is possible—and a very serious problem—for a company to earn handsome net income and run out of cash. In that sense, cash really is king!

All enterprises, both profit seeking and nonprofit, are well advised to do a careful job of cash budgeting and forecasting (see Chapter 9) so as to guard against simply running out of cash. This requires a careful forecast of the timing and amount of cash inflows (from customers, clients, donors, and other sources) and the timing and amount of cash outflows required to assure that, for example, employees (salaries), vendors (discharge of accounts payable), lenders (payment of interest and principal), and shareholders (dividends) remain content—or at least not in open rebellion. The organization needs to decide what "safety stock" of cash it needs to keep; that is, the average minimum balance it should keep in its checking account. If cash inflows are steady and predictable (as is the case for electric utilities and grocery supermarkets), its cash safety stock requirements may be quite low. Incidentally, consciously or unconsciously each of us in our personal lives decides what minimum amount of ready cash to keep on hand.

In this chapter, we take a longer term and more macro look at the flow of cash in and out of an organization. To do so we construct a third financial statement, the **Cash Flow Statement**. All companies whose securities are traded publicly are required to supply this statement at least annually to their shareholders. All other audiences of the financial statements—managers, vendors, lenders, employees—will also be interested in this third statement.

The cash flow statement is derived in major part—that is, constructed—from data available within the balance sheet, the income statement, and the footnotes to those statements. New or additional accounting procedures or processes are not required. One is tempted to ask: why bother, as there is nothing new in this third statement? We bother because (a) the cost of constructing this third statement is minimal, and (b) it highlights some information—about cash—that is hugely important.

THREE KEY SECTIONS OF THE CASH FLOW STATEMENT

Where does cash come from? First, from operations, when cash inflows derived from sales/revenue exceed cash outflows to cover expenses. Remember that this positive difference is not the same as profit, which is calculated on an accrual basis. Moreover, certain expenses on the income statement—primarily depreciation and amortization—do not trigger a cash outflow. Of course, in unsuccessful businesses cash outflows can exceed inflows; then operations are a "user" not a "generator" of cash.

Second, from financing: borrowing more or selling additional common stock. Here, too, the net balance can be an outflow rather than inflow of cash: repayment of borrowing may exceed new borrowing and repurchases of common stock and payment of dividends may exceed cash generated by the sale of newly issued common stock.

A third—and infrequent—source is the sale of long-term assets. More typically, certainly in a growing business, investment in new long-term assets exceeds by a wide margin the sale of such assets. Thus, investments typically do not generate cash; they use cash.

These last three paragraphs define the structure of the cash flow statement. The three key sections are as follows:

- **Cash Flow from Operations**
- **Cash Flow from Investing Activities**
- **Cash Flow from Financing Activities**

The "bottom line" of the statement is the "Net Increase/Decrease in Cash" balances.

FEDERATED'S CASH FLOW STATEMENT

Exhibit 6-1 shows Federated Department Stores' Consolidated Statement of Cash Flows (lightly edited) for the year ended January 29, 2005. Over the course of the year, Federated's cash balance decreased $57 million. Inasmuch as Federated's cash balance at year-end was

Exhibit 6.1: *Federated Department Stores, Inc. Consolidated Statement of Cash Flows Year Ended January 29, 2005 ($ Millions)*

CASH FLOW FROM OPERATING ACTIVITIES

Net income	$689
Depreciation and amortization	743
Change in working capital and other assets and liabilities	75
	1,507

CASH FLOW FROM INVESTING

Purchase of property and equipment	(467)
Capitalized software	(81)
Nonproprietary accounts receivable	(236)
Decrease in notes receivable	30
Disposition of property and equipment	27
	(727)

CASH FLOW FROM FINANCING

Debt issued	186
Debt repaid	(365)
Dividends paid	(93)
Acquisition of treasury stock	(901)
Issuance of common stock	298
Other	38
	(837)
Net Increase/(Decrease) in Cash	($57)

$868 million (see Exhibit 1-2) the decline was just a hair over 6%; Federated still has plenty of cash.

The significance of the statement, then, is not in its bottom line. In summary we see that Federated generated just over $1.5 billion in cash from its operating activities, invested less than half of this ($727 million) in new long-term assets, and utilized $837 million in rearranging its capital structure—that is, to effect changes in long-term borrowing and owners' equity. That all sounds quite healthy—and it is.

Let us delve more deeply into each of these three sections.

Cash Flow from Operations

The top line of the Cash Flow from Operations section is net income. The other entries in this section adjust that number from an "accrual" basis to a "cash" basis. We begin that adjustment by recognizing that some of the expenses—depreciation and amortization—did not consume cash and appear here as if they added cash. Of course, when these long-term assets (fixed and intangible assets) were originally acquired in a previous period, they did consume cash; in those earlier periods the investment in these assets appeared in the second section of the cash flow statement. Some say, inaccurately, that depreciation is a source of cash; in fact the recognition of the decline in value of fixed assets is just a bookkeeping entry and does not involve cash at all. The point is that these noncash expenses did not use cash, as essentially all other expenses do. Think of the sum of the first two lines as "net income before depreciation and amortization."

The next line introduces a new term: **working capital**. Working capital is defined as the difference between current assets and current liabilities. We said earlier that current assets and current liabilities tend to fluctuate with changes in business activity.

A growing firm will typically find that it must increase its working capital each year; that is, the amount by which its accounts receivable and inventories increase (these, you will recall, are typically the dominant current assets) is greater than the amount by which its accounts payable and accrued liabilities (the major current liabilities) increase. That typical increase reduces the cash flow from operations. In fact, at Federated, during the year ended January 29, 2005 this difference decreased by $75 million; it thus adds to, rather than reducing, Federated's cash flow from operations.

An increase in a current asset is a use of cash just as the purchase of a new fixed asset involves a use or outflow of cash. An increase in a current liability is a source of cash just as any increased borrowing provides the corporation with additional cash.

Exhibit 6-2 digs a bit deeper into the year's changes in working capital at Federated (somewhat simplified and abbreviated).

To put this decrease in perspective, note in Exhibit 1-1 that Federated's working capital at January 29, 2005 was $3.2 billion; this decrease then amounts to only 2% of total working capital—not a big deal. Federated's total sales have been essentially flat—no growth or decline—for the past three years and thus we would not expect to see much of a change in working capital.

Note that Federated's cash flow from operations is more than twice its net income. The major reason is that its **noncash expenses** (depreciation and amortization) exceeded its net income this year.

Cash Flow from Investing Activities

The second section of Federated's Cash Flow Statement shows that, despite the absence of growth in sales this year, Federated made significant investments both in property and

Exhibit 6.2: *Federated Department Stores, Inc. Change in Working Capital and Other Assets and Liabilities Year ended January 29, 2005 ($ millions)*

(Increase) decrease in accounts receivable	$17
(Increase) decrease in inventories	95
(Increase) decrease in supplies and prepaid expenses	(5)
Increase (decrease) in accounts payable and accrued liabilities	(24)
Increase (decrease) in current income taxes payable	(6)
Increase (decrease) in deferred income taxes	59
Increase (decrease) in other liabilities	(61)
	$75

equipment and in software. Federated sold a minor amount of property and equipment ($27 million) but bought $467 million. That may sound like a lot, but during this same year additions to accumulated depreciation exceeded that amount; note in Exhibit 6-1 that the sum of depreciation and amortization—a high percentage of this sum is depreciation—was $743 million. Thus, the value of Net Property and Equipment declined this year.

Note the second investment, in Capitalized Software. What does it mean in this context to **capitalize expenditures**? Federated, like most all companies these days, spends a great deal on what is called IT, information technology. Much of this spending is for maintenance, training, updating, and fixing glitches; all of these expenditures are recorded as expenses on the income statement. But some of this spending is to build new and major software systems that will have a lengthy useful life; these systems are long-term assets and thus Federated capitalizes the associated expenditures rather than expensing them.

The investment in Nonproprietary Accounts Receivable is unclear, and unfortunately the footnotes to the financial statement provide only a partial explanation. Apparently, these are amounts due from outside (nonproprietary) credit card companies; proprietary receivables are amounts outstanding on Federated's own credit card systems. The balance sheet shows no value for nonproprietary accounts receivable among the long-term assets and indeed the notes tell us that nonproprietary accounts receivable are included in current assets. Accordingly, I would have expected to see this $236 million investment (increase in receivables balance) in the first rather than the second section of the cash flow statement.

Cash Flow from Financing

The first two entries in this section show the amount of debt repaid and the amount of new debt assumed. The message is that Federated has $179 million (365 minus 186) less long-term borrowing on January 29, 2005 than it did a year earlier. At a company that is not growing but generating substantial operating cash flow, this net debt repayment is what we and Federated's lenders would expect to see.

The remaining entries in this section affect the owners' equity portion of the balance sheet. Recall that dividends are not treated as expenses of the corporation but of course they are paid in cash; this $93 million is shown in parentheses, a use of cash. Federated's net income this year was $689 million (see Exhibit 2-1) and thus dividends paid were about 13% of earnings, a little on the low side for a company that is not growing.

However, the next line on the cash flow statement shows another way in which Federated returned cash to its shareholders: the purchase on the open market of so-called treasury shares to the tune of $901 million—an amount that exceeds by a wide margin the net income for the year. Its annual report shows that Federated devoted over $1 billion over the previous 2 years to repurchase its shares. Let us look a little deeper into these share repurchases.

The notes to the financial statements reveal that Federated bought 18.3 million shares—just over 10% of the shares outstanding at the beginning of the year—for a total of $901 million; thus, the average price paid was $49.23 per share. Given the last year's EPS of $3.93, the average price/earnings (P/E) ratio for these purchases was 12.5, somewhat below the average P/E for all traded common stocks that year. Again, this relatively low P/E is not surprising for a company that is not growing; investors apparently do not find Federated's common stock very exciting. Federated's board understandably decided that the repurchase of its shares was a good use of cash.

At the same time Federated gathered cash in the amount of $298 million from the sale of new common shares. Why would the company sell new shares while it is buying up shares on the market? The very large majority of these sales were the result of employees exercising their stock options. That is, these sales were triggered by the actions of the option holders, not discretionary decisions by Federated. A little more digging into the financial statement footnotes reveals that seven million shares were issued in return for $298 million; the option holders apparently paid an average of $42.57 per share, only about 14% less than Federated paid to purchase treasury stock. The employee option holders certainly did not make a windfall on the exercise of these options.

So, Federated sold seven million shares and bought back 18.3 million, thus reducing the number of outstanding shares by about 12 million (about 1.4%). This reduction in the denominator of the EPS ratio gives a bit of a boost to Federated's EPS.

CASH FLOWS IN OTHER INDUSTRIES

We saw in Chapters 1 and 2 that the industry in which a corporation operates shapes its balance sheet and income statement. This is no less true for the cash flow statement.

Let us explore some company characteristics that will influence its cash flows.

Growth

A company that is growing rapidly will typically be making significant investments both in fixed and intangible long-term assets and also in working capital. Its "cash flow from operations" will be reduced by the working capital investments and its "cash flow from investing activities" will consume substantial amounts of cash. Accordingly, its "cash flow from financing" is likely to show the acquisition of substantial new capital: new borrowings and periodic sales of newly issued common stock. A rapidly growing company, unlike Federated, will struggle to repay debt—indeed its debt is likely to grow each year—and it is very unlikely to pay cash dividends or to repurchase its shares. Indeed, as we will see in Chapter 7, a rapidly growing company may struggle to find sufficient financing to invest in the many high-growth opportunities its industry provides.

Profitability

A new company may incur losses during its early years. Its "cash flow from operations" will likely be negative, while at the same time its "cash flow from investing activities" is likely to include substantial investments in long-term assets: plant and equipment, intellectual property, and so forth. These cash outflows will have to be offset by substantial financing, often primarily equity financing because new and unprofitable companies are generally not very creditworthy.

Federated's profitability is so-so, perhaps what you would expect at a mature company engaged in a business (department store retailing) with limited margins.

Other industries are populated by companies that are hugely profitable: for example, pharmaceuticals, software, investment banking. Many of these companies are in a position to pay handsome dividends, continue aggressive asset investment programs, invest in long-term research programs, acquire other companies for cash, and repurchase their shares.

Asset Intensity

Contrast an electric generating and distributing public utility with a software company like Microsoft. The former has huge investments in fixed assets; the latter requires quite limited fixed asset investments. The former can and must borrow large amounts; the latter has neither the need nor often the opportunity for extensive borrowings. The public utility can generate high cash flows from operations because of its very high depreciation of fixed assets; much of that cash will need to be plowed right back into new fixed assets to service its customers.

In Microsoft's case, combine its very high profit margins with low asset intensity and it is unsurprising that the company has piled up almost obscene amounts of cash—particularly, given the reality that antitrust considerations will probably preclude it from making any large corporate acquisition. Finally, a few years ago Microsoft declared an enormous one-time cash dividend and began paying regular annual dividends. Arguably, directors have a fiduciary responsibility to return any excess cash to its owners, the shareholders. Federated is doing this both by paying cash dividends and by repurchasing shares.

NEW TERMS

Capitalizing (expenditures)	Cash flow statement
Cash flow from financing activities	Noncash expenses
Cash flow from investing activities	Working capital
Cash flow from operations	

CHAPTER 7

Evaluating with Ratios
The Key to Analysis

Recall the definition of accounting at the beginning of Chapter 3. We have now completed the "observing, measuring, recording, classifying, summarizing" parts of that definition—and we have reviewed the three financial statements where the summarized data are *presented*. In this chapter, our task is to analyze those three financial statements, attempting to draw out as much information as possible to help make better decisions—whether we are managing the business, thinking about investing in it, contemplating lending to it, or considering it as a customer, supplier, or employer.

Earlier chapters have hinted that individual dollar amounts appearing on the balance sheet, income statement, or cash flow statement do not tell us much. If we know the net income for Corporation F for last year, so what? We need to relate that piece of data to something else—perhaps to total revenue, to total assets, or to total owners' equity—in order to be able to evaluate F's profitability. If we know that F's total accounts receivable at the end of the year were $3.1 million, so what? If, however, we also know that F's total sales last year were $25 million, we can (as we will see in this chapter) estimate the average time it takes F's customers to pay their bills.

So, **ratios** are the key to drawing information from financial statements: ratios involving amounts within a single statement and ratios involving amounts drawn from two different statements.

RATIO ANALYSIS

In addition to the conventional ratios that we will concentrate on in this chapter, analysts are free to invent new ratios that just happen to fit the business or the circumstances. For example, since Federated borrows substantial amounts, both short term and long term, a ratio of the two might be useful; although that is not a ratio commonly used.

Calculating ratios is somewhat analogous to performing a medical examination on a patient. Both the financial analyst and the medical doctor are looking for signs of strength and signs of weakness or disease. Very often these analyses—financial ratios or medical diagnostics—

suggest other tests or investigations to yield more useful information. And not every ratio is useful in every situation.

Account values bounce around a good bit, financial period to financial period—what electrical engineers call noise in the system. That is, not every small shift in dollar amounts is significant. Accordingly, there is no point in calculating ratios to three or four significant figures; two are typically plenty.

With practice you will learn to calculate ratios in your head, rather than on a calculator. You will not read financial statements as you would a book, from left to right, top to bottom of the page. Instead, you will relate various values as you peruse the statements.

Categories of Ratios

While there are endless ratios, we can classify them into about five categories:

- Liquidity
- Utilization of working capital
- Capital structure
- Profitability
- Cash adequacy

This chapter illustrates these ratios using Federated's financial statements that appear in Chapters 1, 2 and 6.

LIQUIDITY RATIOS

The question here is how able the corporation is to meet its near-term obligations: pay salaries, discharge accounts payable, pay interest on borrowing, and so forth.

Current Ratio

The **current ratio** is kind of the granddaddy of all ratios: current assets divided by current liabilities. Recalling the definitions of these two subtotals on the balance sheet, you will recognize that they are defined in a parallel manner: assets that are or will become cash within the next 12 months and obligations (liabilities) that must be discharged within the next 12 months.

Do not fall into the trap of thinking that current assets include all sources of cash that will be received in the next 12 months, nor that liabilities show all claims on cash for the next 12 months. Salaries that will be earned next month do not appear in current liabilities and yet they will have to be paid in cash either next month or the month after. Similarly, the company's sales next month are not yet in A/R on the balance sheet, but surely the customer will pay his or her bill within the next 12 months.

Federated's current ratio at January 29, 2005 was $7,510/4,301 = 1.7$

Quick or Acid-Test Ratio

Another liquidity ratio relates just the sum of the most liquid of the liquid assets—cash, cash equivalents, plus accounts receivable (so-called quick assets)—to total current liabilities.

Federated's **quick ratio** at the end of the year was $(868 + 3,418)/4,301 = 1.0$.

Note that the denominator of these two liquidity ratios is the same; the primary difference in the numerator is that inventory is not included in quick assets.

At one time conventional wisdom held that adequate liquidity demanded a current ratio of 2.0 or higher. Beware of conventional wisdoms! Many corporations—certainly including Federated—can comfortably operate with current ratios well below 2.0; and, if the company had no inventory (for example, a service company) or no accounts receivable (all cash transactions), a current ratio below 1.0 might be satisfactory. We can conclude here that Federated is comfortably liquid.

UTILIZATION OF WORKING CAPITAL RATIOS

Recall that working capital—current assets and current liabilities—arise because of the active conduct of business and could shrink to zero if operations ceased. Still, investments in working capital are often very large. This set of ratios tests whether the components of working capital are being used efficiently. This idea will come clear as we discuss the individual ratios.

Accounts Receivable Collection Period

This ratio is (accounts receivable)/(average sales per day), where "average sales per day" is simply annual sales divided by 365. This ratio tells the analyst the average number of days customers take to pay their bills. The units of the **accounts receivable (A/R) collection period** are "days": the number of days worth of sales represented by the year-end receivables balance.

For Federated, the calculation is $3,418/(15,630/365) = 80$ days. A couple of cautionary comments about this ratio:

- What are the corporation's normal terms of sale? Most business-to-business transactions are "net 30 days." But Federated's are business-to-consumer transactions and the various Federated store groups offer extended credit terms to their customers.

- What is the mix of "cash" and "credit" sales? Again, business-to-business transactions are almost always credit; not so at Federated, but we do not know the mix. Surely, if the denominator was limited to average *credit* sales per day, the collection period would be longer than 80 days.

- Does the company experience seasonal swings? Certainly Federated does; its annual report states that a high proportion of its sales occur in November and December. Indeed, this seasonality is almost certainly the reason why Federated chose to end its fiscal year at the end of January—comfortably past the busy season—not at December 31. Had we calculated Federated's collection period at December 31, the likelihood is that its "days of receivables" would have been well above 80.

- Is the corporation growing or shrinking rapidly? Receivables outstanding at year-end are a function of the rate of sales (sales per day) at the end of the year. Customer receivables from the beginning of the year have been collected by the end of the year. Thus, if a corporation is growing at, say, 15 or more percent per year, consider adjusting the collection period for this increase in sales rate. A simple adjustment is to average the beginning and end-of-year accounts receivable balance. The ratio then becomes:

$$\text{Average A/R collection period} = (\text{Avg. A/R})/(\text{Annual sales}/365).$$

We are hard pressed to discern whether 80 days is a reasonable collection period for Federated. We would be in a better position if we knew Federated's collection period in earlier years or the collection period of some of Federated's competitors (other department store groups). We will return to those thoughts later in this chapter.

Inventory Turnover

Just as the ratio of sales to accounts receivable is useful, a similar ratio involving cost of goods sold and inventory is useful. In the former case, both numerator and denominator are valued at "sales prices"; in the latter case, both are valued at "acquisition cost."

A ratio called the **Inventory Flow Period** is exactly analogous to the A/R collection period (again, measured in days):

$$\text{Inventory}/(\text{Annual cost of goods sold}/365).$$

People tend to think of **inventory turnover** in times per year, rather than number of days. That ratio, measured in times per year is simply

$$(\text{Annual cost of goods sold})/\text{Inventory}.$$

Federated's ratios for the most recent year are

$$\text{Inventory Flow Period} = 3{,}120/(9{,}297/365) = 122 \text{ days}$$
$$\text{Inventory Turnover} = 9{,}297/3{,}120 = 3.0 \text{ times per year.}$$

Two of the cautions relating to A/R collection apply also to inventory turnover:

- You can be sure that Federated's inventory was a great deal higher on November 15, 2004, just before the busy season, than it was on January 29, 2005. Perhaps it will be still lower at the end of March, after Federated has completed its winter clearance sales.

- Is the company changing size rapidly—growing or shrinking? Federated's activity has been, in fact, pretty flat for the past 3 years.

And, again, it would be useful to compare Federated's inventory turnover data to its competitors', and to review the trend in these ratios over the past several years.

Accounts Payable Payment Period

Now turning to the current liabilities side, we can ask analogous questions about accounts payable:

A/P payment period = accounts payable/(annual credit purchases/365).

Neither the statements nor the footnotes reveal what Federated's total credit purchases were for the year. Let us look for a reasonable proxy, that is, a readily available financial value that typically moves in concert with credit purchases. We can safely assume that Federated purchases all of its merchandise inventory on credit. Federated also purchases many supplies and services (utilities, advertising, and so forth) on credit, but these would be a small fraction of its merchandise credit purchases. Lacking information on what aggregate credit purchases were for the year, assume that cost of sales of merchandise is a reasonable proxy.

Federated's balance sheet shown in Exhibit 1-1 unfortunately combines accounts payable and accrued liabilities, $2,707 million. Fortunately, the notes to the financial statements show "accounts payable and expense accounts payable," $1,301 million. Using this figure, the accounts payable payment period at year-end was

$$1,301/(9,297/365) = 51 \text{ days.}$$

We know that this calculation still overstates the payment period because the $1,301 million overstates the accounts payable attributable solely to merchandise purchases. But, if we knew what this ratio was for each of the last several years, we could estimate whether Federated is changing the timing of its payments to vendors. That is, the use of a proxy causes the *absolute* value of this ratio to be of questionable value; nevertheless, if the proxy remains unchanged, the *trend* information is useful.

Once again, as was the case for the A/R collection period and inventory turnover ratios, we need to be alert to seasonality and growth/decline effects on this ratio. And, again, it would be helpful to look at the trend in this payment period for the last several years before drawing any conclusions about the timeliness of Federated's payments to its vendors.

An interesting question: are companies well advised to maintain a low accounts payable payment period? The company's vendors will be pleased, but is it not sometimes wise to "use" the vendors' money, a source of short-term borrowing, since vendors seldom charge interest for this borrowing? Yes, but be careful. As mentioned in Chapter 5, vendors may lose patience with customers who pay slowly; they may begin to provide poorer service or uncompetitive prices—and finally may decide to stop doing business with the slow payer. That decision can have dire consequences if that vendor provides parts, merchandise, or services that cannot be obtained readily elsewhere.

We will sneak one more ratio in here that does not really qualify as a working capital ratio, but one that will prove useful later: the **total asset turnover ratio**. This is the ratio of total annual sales to total assets. This ratio indicates how much a company must invest in order to generate a dollar of annual sales. It is a gross measure of asset productivity. For Federated in the year ended January 29, 2005, total asset turnover was

$$15,630/14,885 = 1.05 \text{ times per year.}$$

This asset productivity appears quite low for a merchandising operation, more like one we would expect to see in a complicated manufacturing operation. Part of this may have to do with rather high investment in property and equipment. Once again, comparing this ratio with that of other companies in the department store industry might be revealing. When calculating this ratio for companies having large investments not used in their primary business activities—for example, excess cash balances or passive equity investments in other corporations—the value of these unrelated assets should be deducted.

Why is this asset turnover ratio important? Because, of course, these assets must be financed, and the capital to finance them has costs. More on this in a few moments.

CAPITAL STRUCTURE RATIOS

These ratios focus only on the "right-hand side" of the balance sheet: liabilities and owners' equity. The question before us is: how is this company financed? To what extent does it use debt? Recall that debt can have both positive and negative leverage effects, but high debt leverage adds risk to the business.

There are many possible capital structure ratios but the "information content" of many of them is the same. Two are as follows:

Total debt/total assets
Long-term debt/total capitalization

Federated's total debt to total assets ratio on January 29, 2005 was

$$8,718/14,855 = 0.59.$$

Total capitalization in this context (this term "capitalization" is used in many different ways!) is defined as long-term debt plus owners' equity. The next question is whether we should include Deferred Income Taxes as part of long-term debt. If the company believes that these taxes will be deferred indefinitely, they should be omitted. Under that assumption, in the case of Federated

$$\text{Long-term debt/total capitalization} = 3,218/9,385 = 0.34.$$

So, we can say that long-term debt provides 34% of Federated's "permanent" capital (omitting current liabilities) and owners' equity provides 66%.

These ratios suggest that Federated is not particularly debt leveraged. Recall that Federated now faces substantial cash demands to effect its acquisition of The May Department Store Company. Borrowing to meet that need could substantially increase Federated's debt leverage and therefore its financing risk. In fact, 9 months later, following the acquisition and despite substantial increases in owners' equity (because of the large issuance of new common stock to May's shareholders), the long-term debt to total capitalization ratio had increased to 44%.

Federated's debt leverage—both before and after the May Company acquisition—seems not to signal undue financing risk for the company. It is not unusual for mature, stable companies to carry long-term debt equivalent to more than half of their total capitalization.

Some smaller companies borrow short term for permanent capital needs, typically because their weak credit ratings do not permit them to borrow long term. In these situations, the best leverage ratio may be

Total liabilities/owners equity

or

Total liabilities/total assets.

Another way to evaluate the use of debt in the capital structure is in terms of the borrower's ability to pay the interest on the debt. Thus, another useful ratio is

Times interest earned = (Income before interest and taxes)/annual interest expense.

This ratio is sometimes called the **interest coverage ratio**. For Federated that ratio is

$$1,415/299 = 4.7 \text{ times.}$$

This coverage is very comfortable.

PROFITABILITY RATIOS

As mentioned in earlier chapters, we evaluate profitability as it relates to both sales and assets; that is, how profitable the company is per dollar of sales, and how profitable the company is in relation to the funds invested in the business.

Running percentages all the way down the income statement yields useful information. Net income divided by net sales is referred to as the **return on sales (ROS) ratio**. The percentages for Federated in the year ended January 29, 2005 were as follows:

Net sales	100.0%
Cost of sales	59.5
Gross margin	40.5
Selling, general & admin. expenses	31.5
Operating income	9.0
Interest expense	(1.9)
Interest income	–
Income before income taxes	7.1
Income tax expense	(2.7)
Net income	4.4%

Federated's gross margin, operating income, and net income percentages are somewhat below what one would expect for a manufacturing company, but higher than what a grocery supermarket chain would typically generate. Analysts seeking a better understanding of Federated need to compare these percentages to those at other department store groups, and to look at changes in these percentages over time.

The dominant ratio to measure of return on investment is the **return on equity (ROE)**: net income divided by total owners' equity. This ratio tells what the rate of annual earnings is as a percent of the funds entrusted to the company by its shareholders. But typically the book value of owners' equity and the aggregate market value of the outstanding shares (defined as "market capitalization" in Chapter 5) are quite different values, sometimes higher, frequently lower. What investors are most interested in is the ratio of net income to market (not book) value of equity. Nevertheless, this ROE ratio is useful for comparing companies across different industries.

Federated's ROE is 689/6,167 = 11.2%. For American corporations these days, this ratio is OK, but not great. Average ROEs are in the mid-teen percentages (of course ROEs are affected by economic cycles). Does an 11% return provide shareholders adequate return for the risk they take in investing in Federated? Perhaps the directors feel that it does not and it therefore authorized in early 2005 the acquisition of The May Department Stores. As a separate company May has lower ROS than Federated, but perhaps Federated's board believes that economies of scale achievable by the combination, coupled with the higher debt leverage (remember, Federated will borrow a substantial portion of the acquisition cost) will improve the combined corporations' ROE.

Recall from Chapter 5 that ROE is affected by debt leverage: the higher the debt leverage, the more ROE will be leveraged up when company performance is strong and leveraged down when performance is weak. Thus, a company may be able to improve its ROE by borrowing more of its total required capital (and thus relying less on the sale of common stock). But the corollary is that the company thereby assumes more risk. Thus, the ROE profitability ratio must be interpreted in light of the corporation's debt leverage position.

Another useful profitability ratio is one that factors out the effects of leverage, and measures profitability as a percent of all assets regardless of how the company chose to finance those assets: the **return on assets (ROA)** ratio is

$$\underline{\text{Return on assets}} = \underline{\text{earnings before interest and taxes/total assets.}}$$

Note that "earnings before interest" are unaffected by the amount of the company's borrowing. And, because interest is deductible for income taxes but dividends are not, this ratio omits income taxes as well.

Federated's ROA was 1,400/14,885 = 9.4%. When comparing ROA and ROE, remember that the former is pretax and the latter is after income taxes. Fair to say that Federated's ROA is decidedly ho-hum!

CASH ADEQUACY RATIOS

Turn now to a set of ratios to probe another question: does the company generate adequate cash? Its needs include cash to pay dividends, acquire other companies, retire borrowing on schedule, repurchase its common stock, invest in capacity expanding or cost reducing capital equipment, and so forth. Two ratios discussed above are aimed at the same question:

The current ratio: how able is the company to meet its near-term obligations?
Times interest earned: how much of a cushion does the company have in meeting its interest obligations (since failure to pay interest on time can be catastrophic for the company)?

In fact, a more useful ratio would look at cash flows, not operating earnings, drawing information from the cash flow statement as well as from the balance sheet and income statement. Here are some possibilities:

a) Cash flows from operating activities/Net income.
b) Dividends/Net cash flows before financing.
c) Debt service requirements/Cash flows before financing.
d) Cash flows from operating activities/Cash flows for investments.

Most companies, because of noncash expenses, have "cash flows from operating activities divided by net income" of greater than 1. But how much greater? A company with very heavy investments in depreciable or amortizable assets is likely to have a ratio considerably higher than 1. Another company that is working hard to improve its A/R collections and inventory turnover may be able to generate cash flows much above its net income, as it reduces its working capital investments. On the other hand, a rapidly growing company will be investing in working capital and thus its cash flows from operations may be less than net income.

The second and third of these four ratios are coverage ratios: (b) dividends coverage, since cash flows from operations less cash used for investments are available to pay dividends; (c) debt service coverage where debt service is the sum of interest payments and principal repayments that the company was obligated to pay during the year. These ratios may help (b) an investor decide whether a corporation might increase its dividend rate (if this ratio is low) or whether the current high dividend rate might be in jeopardy (as this ratio approaches or exceeds 1), and (c) a lender decide whether a company can comfortably meet the terms of its borrowing agreements.

Another coverage ratio is the **dividend payout ratio**: dividends per share (DPS) divided by earnings per share (EPS). This ratio tells you what percentage of current earnings is being returned to the shareholders as dividends; thus, it too is a kind of "dividends coverage" ratio, but inverted. And, while we are on the subject, another ratio important to shareholders is **yield**: annual dividend payments per share divided by market price per share, generally quoted as a percentage. Note that if dividend payments remain unchanged and the market price of the stock declines, the stock's yield increases—analogous to the relationship between bond yields and bond prices. And the reverse: an increase in the dividend rate may result in a higher price for the stock if investors bid up that price until yield returns to its former percentage.

INTERPRETING RATIOS

Interpreting ratios is an art, not a science. To repeat, the analyst is using ratios to diagnose corporate strengths and weaknesses. Ratios are typically not the end of the analysis but the beginning; they help guide the analyst to those areas and aspects of the business where he or she wants to dig deeper, ask more questions.

Despite tempting conventional wisdoms, there are no "right values" for various ratios. In fact, you cannot even say that the higher the ratio—or the lower the ratio—the better. Just because very high debt leverage is very risky does not mean that very low debt leverage is a good thing; is the corporation that eschews debt wise to forego the positive effects of debt leverage? Again, while low liquidity ratios (for example, a low current or quick ratio) may signal danger, a very high current ratio signals inefficient use of current assets—perhaps excess cash

Exhibit 7.1: *Federated Department Stores, Inc. Key Financial Ratios, 2002–2004*

	END OF JANUARY		
RATIO	2005	2004	2003
Current ratio	1.7	1.9	
Accounts receivable collection period (days)	80	72	
Inventory turnover (times per year)	3.0	2.8	
Total asset turnover	1.05	1.05	
Long-term debt/total capitalization (%)	34	39	
Gross margin	40.5%	40.4%	40.0%
ROS (return on sales)	4.4%	4.5%	4.1%
ROE (return on equity)	11.2%	11.7%	
ROA (return on assets)	9.4%	9.2%	
Cash flow before financing/dividends (times)	8.4	14.9	Inf.
Cash flow from operations/cash flow for investing	2.1	2.4	1.8

or too much inventory. The corporation's finance officer may seek rapid A/R collection, but to the marketing manager "payment terms" are only one element of a complex customer-supplier relationship; the optimum collection period trades off financial and marketing considerations.

As suggested several times in this chapter, analysis of trends and comparisons with similar companies can be particularly illuminating.

Trends

Exhibit 7-1 shows some key ratios for Federated over 2- and 3-year periods. Consider whether these provide further insights into Federated's performance or financial position.

Note that the ratio values for 2005 are those discussed above.

What observations (if not conclusions) can we draw from Exhibit 7-1? As said repeatedly in this chapter, Federated is a mature and stable business; we see no great swings or changes in these ratios. Nor, frankly, do we see any significant improvements. Federated's performance is somewhat unexciting and its financial position is solid. We can imagine that Federated's board is feeling pressure to take bold action to improve performance, pressure that probably led to the May Department Stores acquisition agreement in February 2005.

Exhibit 7.2: *Ratio Comparisons, Federated and Target*

	FEDERATED	TARGET
Accounts receivable collection period	80 days	39 days
Inventory turnover (times per year)	3.0	5.9
Total asset turnover	1.05	1.45
Long-term debt/total capitalization	34%	41%
Gross margin	40.5%	32.9%
Return on sales (ROS)	4.4%	6.8%
Return on equity (ROE)	11.2%	24.5%

The decline in current ratio is modest, and probably not particularly significant. While A/R collection has declined by about 10% in the most recent year, inventory turnover has improved. Gross margin is creeping up as is ROS. The small changes in ROE and ROA—one up slightly, the other down slightly—are probably not significant. The "cash flow before financing/dividends" tells us not much about performance, but does suggest a change in dividend policy; Federated paid no dividends in 2003, and increased dividends pretty smartly between 2004 and 2005.

Comparisons within the Department Store Industry

Comparing the values of Federated's key ratios with averages for the department store industry overall would surely be useful, as would "benchmarking" Federated's ratios with those of its most successful competitors. Exhibit 7-2 compares Federated with another and larger retailer, Target Corporation. While one might argue that the two companies are not in exactly the same business, Target is relatively younger, has had recent success in growing sales and earnings, and is now a formidable retail competitor; accordingly, Target is a useful benchmark for Federated.

Target's primary advantages over Federated are better utilization of working capital and lower operating expenses; despite a lower gross margin, Target's ROS is 150 percent of Federated's. With only slightly more debt leverage, but with rapid total asset turnover, Target is able to rack up an ROE more than twice that of Federated. Further, while Federated's growth over the past 4 years has been nil, Target has grown sales at a 9% compound rate and EPS at 16%.

SUSTAINABLE GROWTH RATE

Surely by now it has occurred to you that examining some ratios in combination might yield some interesting information. Or, put another way, are there some interesting linkages among these ratios? The linkage among (a) ROS, (b) efficiency of asset utilization, and (c) debt leverage is particularly revealing. For example, a company with very high ROS may be able to invest heavily in assets without incurring excessive debt. Another company with high tolerance for debt and very efficient use of assets may be able to sacrifice some return on sales, perhaps in order to gain market share against its competitors. And a third with low tolerance for debt and only average ROS may need to hold back on its investments in new assets, and thus moderate its growth rate. And we could go on!

An interesting question that should interest managers, lenders, and investors alike is this: how fast can a particular company grow its revenues/sales without having to either sell additional equity capital or increase its debt leverage? This question is particularly pressing for relatively young (immature) companies with high growth aspirations.

In a moment we will ask that question about Federated. But first, consider hypothetical company H that has the following ratios:

ROS	4%
ROE	15%
Total debt/total assets	60%
Total asset turnover	120%

(All of these ratios were discussed above.)

Seeing lots of opportunity in its major markets, Company H is interested in pursuing rapid sales growth, say 25% per year, so as to capture emerging opportunities and maintain or increase its market share. Assume that in the year just completed, H had net sales of $24 million. If it achieves 25% growth, its sales next year will grow to $30 million. By how much will its assets need to grow to support that sales growth? Well, if H's total asset turnover remains unchanged, it will need to grow its total assets by 25%. That is, if its A/R collection period and inventory turnover remain the same and H continues to invest in fixed assets at the rate it has been investing, its assets by the end of next year will have grown by 25%. This simple projection assumes that H will be no more or less efficient in its use of assets next year than it was this year (a reasonable first assumption).

Recall the accounting equation: Assets = Liabilities + Owners' Equity. If assets are to increase by 25%, then so also must the right-hand side of the equation, the sum of liabilities and owners' equity. Two more reasonable assumptions: H would like to avoid any additional equity financing in the coming year, and it plans to continue its practice of paying no cash dividends. Given these assumptions, by how much will owners' equity grow next year? Invested capital will not grow at all, and retained earnings will grow by the amount of profit H earns, all of which will be retained. If H's ROE ratio next year is again 15%, retained earnings will grow by 15% of total owners' equity.

Back to the accounting equation. If the left-hand side grows by 25% and a portion of the right-hand side (owners' equity) grows by only 15%, then the other portion of the right-hand side—liabilities—will have to make up for the short fall. It will have to grow by more than 30%. Now, that boost in debt leverage (total debt to total assets will grow from 60% to over 63%) may be quite achievable and tolerable in the coming year. But debt leverage cannot increase year after year without at some point pushing the riskiness of the business too high-if not too high for H's owners and managers, then too high for its lenders!

Thus, ROE is the governor on how fast a company can grow on a sustainable basis without doing additional equity financing. We can illustrate this phenomenon as follows:

$$\text{ROE} = (\text{net income/sales}) \times (\text{sales/assets}) \times (\text{assets/owners' equity}).$$

The first of these ratios is ROS, the second is asset turnover, and the third is a measure of debt leverage. The third ratio is not one discussed above, but note that the difference between the numerator and the denominator (assets minus owners' equity) is "liabilities," and the greater this difference, the greater the debt leverage.

Look again at this last equation and you will see that both "sales" and "assets" cancel out and we are left with "net income/owners' equity," the definition of ROE. The point of the equation is to call attention to the linkages: ROE is simply the product of ROS, asset turnover, and debt leverage. Company H presumably needs to increase its ROE to something like 25% if it wishes to sustain an annual revenue growth of 25% and—this is an important "and"—sell no additional common shares. It can do that by increasing its debt leverage, but only up to a point. H's managers need to think about how they can increase either or both of ROS and asset turnover to pump up the company's ROE. Or, if competitive pressures seem to preclude those actions, H better get prepared to do additional equity financing or plan for slower growth.

Let us check the sustainable growth situation at Federated using the last equation:

$$\text{ROE} = 4.4 \times 1.05 \times 2.41 = 11.13\%.$$

For the past three years, Federated has not grown. Thus, as we saw on Federated's cash flow statement, the company could and did use its profits to pay dividends and repurchase its common stock.

A reminder: companies can only invest their *retained* income not their total income. Thus, the ROE must be factored by the dividend payout rate when analyzing the sustainable growth rate of a company: $(1 - \text{DPS/EPS})\text{ROE}$. This calculation underscores why high-growth companies are reluctant to pay dividends: they need all their earnings to invest in growth opportunities.

ANOTHER EXAMPLE: ANALYSIS OF A MANUFACTURING COMPANY

Let us analyze another company that operates in a very different industry than the department store world: manufacturing a technically sophisticated line of products. The Wellstock Corporation is a disguised, small company that provides design, production, and support of communications and aviation electronics products for commercial and military customers worldwide. Its common stock is publicly traded and the company enjoys a strong reputation in its industry. (Wellstock's financial statements have been somewhat simplified and clarified for presentation here.)

Before we look at detailed financial statements, consider some of the key differences between this commercial-military manufacturing industry, with a strong emphasis on technology, and the department store industry in which Federated competes, and how these differences will be reflected on the financial statements. Here are some of those differences:

- The inventory at Wellstock arrives as raw material and parts and is fabricated, assembled into complex systems, and then shipped to customers; Federated's inventory arrives and is shipped to customers essentially unaltered.

- Wellstock's plant and equipment is very different from Federated's.

- One imagines that Wellstock's gross margin and ROS will be somewhat higher than Federated's; on the other hand, we have no basis for speculating that their ROEs will be substantially different.

- Wellstock's terms of sales are probably "net 30 days" but, on the other hand, shipment, installation, and testing times are likely to push out the date when the customer will agree to make payment; accounts receivable collection periods are likely to be relatively long, although probably not longer than Federated's.

- Wellstock is likely to hold significant raw and in-process inventory but little or no finished goods inventory (except spare parts held for servicing).

- Given the technology risk assumed by Wellstock and the unpredictability of dealing with government customers, Wellstock may feel that it cannot tolerate high debt leverage.

- Warranty and service work by Wellstock staff is likely to be important.

- Without knowing more about its products and customers, it is difficult to speculate as to how much Wellstock needs to spend on marketing.

- If Wellstock is growing rapidly, it may pay out little or none of its earnings in dividends.

With those predictions, let us see what Wellstock's financial statements look like for the fiscal year ended September 30, 2005. Exhibits 7-3, 7-4, and 7-5 present, respectively, what Wellstock refers to as "Consolidated Statement of Financial Position" (balance sheet), "Consolidated Statement of Operations" (income statement), "Consolidated Statement of Cash Flows" (cash flow statement).

What is remarkable on this balance sheet (Exhibit 7-3)? The current ratio is 1.5; adequate but not overly liquid. With cash less than 10% of current assets, Wellstock does not appear to have excess cash. Total debt-to-equity is rather high at 234%, but long-term debt is only 18% of total capitalization and we can probably assume that the retirement benefits, by far the largest liability, are payable over many years (that is, will not be a near-term call on cash). Note the warranty reserve in current liabilities; this liability will be discharged, presumably, by performance of services rather than by payout of cash. Note the substantial balance in treasury stock (last line item in shareowners' equity); the cash flow statement should tell us more.

Turning then to Exhibit 7-4, note that Wellstock has helpfully separated both its sales and cost of sales accounts into those relating to products and those relating to services. This separation permits two gross margin calculations: on products, 27%; on services, which account for only about 11% of total sales, 30%.

Selling, general and administrative expenses are surprisingly low at only 11.6% of sales. Engineering costs, particularly on military products, are probably included in cost of sales of products. The low selling costs may be a result of Wellstock having relatively few customers. Note that both "basic" and "diluted" earnings per share are given, the difference being that the effect of outstanding stock options is included in the denominator of the "diluted" EPS calculation. Dividends paid (DPS) amount to only 22% of the diluted earnings per share.

We can now calculate the several useful ratios that draw data from both the balance sheet and the income statement:

A/R collection period	80 days
Inventory flow period (product cost of sales)	112 days
Total asset turnover	1.1 times
Return on equity (ROE)	42%
Return on assets (ROA)	17.6%

Exhibit 7.3: *Wellstock Corporation Consolidated Statement of Financial Position September 30, 2005 ($ millions)*

ASSETS

Current assets	
Cash and cash equivalents	$15
Receivables	73
Inventories	69
Current deferred income taxes	18
Other current assets	3
Total current assets	178
Property	47
Intangible assets	11
Goodwill	46
Other assets	32
Total assets	$314

LIABILITIES AND SHAREOWNERS' EQUITY

Current liabilities	
Accounts payable	$29
Compensation and benefits	27
Income taxes payable	4
Product warranty costs	17
Other current liabilities	40
Total current liabilities	117
Long-term debt	20
Retirement benefits	76
Other liabilities	7
Shareowners' equity	
Invested capital	126
Retained earnings	17
Treasury stock	(49)
Total shareowners' equity	94
Total liabilities and shareowners' equity	$314

Exhibit 7.4: *Wellstock Corporation Consolidated Statement of Operations Year Ended September 30, 2005 ($ millions)*

Sales	
Product sales	$307
Service sales	37
Total sales	344
Costs, Expenses and Other	
Product cost of sales	224
Service cost of sales	26
Selling, general and administrative	40
Interest income and expense	(1)
Total costs, expenses and other	289
Income Before Income Taxes	55
Income Tax Provision	15
Net Income	$40
Earnings per Share, Basic	$2.24
Earnings per Share, Diluted	$2.20
Cash Dividends per Share	$0.48

Both the A/R collection period and the inventory flow period appear to be long—about the same as Federated's—but that may simply be a consequence of the business that Wellstock is in. Given those ratios, it is not surprising that the total asset turnover is only 1.1 times.

Without question, the ROE is very handsome indeed. Recall that Wellstock has bought up a good deal of its own common stock and thus the low value of the denominator of the ROE ratio is a major factor.

Look at Exhibit 7-5, Wellstock's cash flow statement for 2005. Operating activities brought $59 million in cash to Wellstock and only $13 million of that, 22%, was spent on investments. The major news in the cash flow statement is that $59 million was returned to shareholders through a combination of dividends and share repurchases. The cash from the exercise of stock options by employees, $10 million, is not inconsequential in the cash flows of the company. Finally, the $5 million decrease in cash balances (including cash equivalents) represents 25% of the cash balance at the beginning of 2005, but Wellstock still appears to have adequate liquidity.

Exhibit 7.5: *Wellstock Corporation Consolidated Statement of Cash Flows Year Ended September 30, 2005 ($ millions)*

Operating Activities	
Net income	$40
Depreciation and amortization	12
Pension plan contributions	(11)
Compensation and benefits paid in common stock	7
Decrease (increase) in working capital	8
	57
Investing Activities	
Property additions	(11)
Acquisition of businesses	(2)
Other	–
	(13)
Financing Activities	
Purchases of treasury stock	(50)
Proceeds from the exercise of stock options	10
Cash dividends	(9)
	(49)
Net Change in Cash and Cash Equivalents	($5)

Wellstock's 2005 annual report provides 5 years of historical financial information, fiscal years 2001 through 2005, that permits a review of trends in some key dollar amounts and ratios. Exhibit 7-6 shows that information.

These financial data and ratios permit us to make some observations, draw a few conclusions, and frame questions that we might like to investigate were we to pursue a more comprehensive analysis of Wellstock. Remember that much that has occurred at Wellstock and in the markets and economic environment in which it competes is not contained in Wellstock's financial records. It is for that reason that most observations and conclusions can, at this stage, be only tentative.

Not all will agree on the significance of the financial trends revealed in Exhibit 7-6, but here are my thoughts:

- Growth. With 5 years of data we can calculate approximate compounded growth rates for some of the key data from 2001 through 2005. Sales have grown only about 5%

Exhibit 7.6: *Wellstock Corporation Selected Financial Data and Ratios, 2001–2005*

FINANCIAL DATA AND RATIOS	2005	2004	2003	2002	2001
Sales ($ millions)	$344	$293	$254	$249	$282
Gross margin (%)	27.4	26.8	26.6	25.2	25.2
Selling, gen'l & admin. expense (% of sales)	11.7	12.2	13.4	12.3	12.4
Income tax rate (% of pretax income)	27.1	30.0	29.9	30.8	37.9
Return on sales—ROS (% of sales)	11.5	10.3	10.1	9.5	4.9
Return on Equity—ROE (% of ending equity)	42.2	26.6	31.0	23.9	12.5
Debt (long term) to equity (%)	21.3	17.7	5.0	13.4	18.2
Asset turnover (times per year)	1.1	1.0	1.0	1.0	1.1
Working capital turnover (times per year)	5.8	4.2	4.8	6.2	5.6
Fixed asset turnover (times per year)	7.3	7.0	6.3	6.1	6.4
Capital expenditures as percent of net income plus depreciation and amortization	22	22	19	18	42
Per-Share Data:					
Earnings per share (EPS)—diluted ($)	2.20	1.62	1.43	1.28	0.80
Dividends per share (DPS) ($)	0.48	0.39	0.36	0.36	0.09
Median stock price ($ per share)	42.10	31.63	22.44	20.35	18.02
Price earnings ratio (P/E)	19.1	19.5	15.7	15.9	22.5
Yield (%)	1.1	1.2	1.6	1.8	–
Payout ratio (DPS/EPS, %)	22	24	25	28	11
Total capitalization/equity book value (approx.)	8.1	5.2	4.9	3.8	2.8

per year, but earnings per share have grown about 29% per year and Wellstock's stock's market price has grown 24% per year. (Wellstock's earnings may have been abnormally low in 2001; the EPS growth rate over the last 3 years has been 20% compared to 29% over 4 years. This difference emphasizes that growth rates can be much distorted if the first or last year is abnormal.) Investors who have owned this stock over the 4-year period should be quite pleased!

- If sales growth has been modest, but profit growth handsome, other phases of Wellstock's operations must have performed well. Note that the gross margin has increased by about two percentage points and operating expenses (selling, general and administrative) have been reduced somewhat. The combined effect is about a 3% increase in profit before taxes—good, but not spectacular. Importantly, Wellstock's income tax rate has decreased by almost 11% over these years, a major contributor to the 2% ROS improvement over the last 3 years. A question worth investigating further: what are the causes of this lower tax rate and is it sustainable in the years ahead?

- With its strong positive cash flow, Wellstock might have augmented its cash balance rather than using the cash to purchase treasury shares and pay (modest) dividends. We can reasonably ask why Wellstock does not have more high-return investment projects to which it could devote this extra cash. On the other hand, in the absence of such opportunities, the company can be commended for returning the cash (in two ways) to its shareholders rather than simply fattening its bank account. At $15 million, less than 5% of total assets and just over 5% of annual sales, Wellstock's cash (cash and cash equivalents) is modest; it actually declined by $5 million last year.

- Wellstock earned a remarkable 42% ROE in 2005, a substantial increase and well above industry averages in 2005. We see that debt leverage has increased substantially—by a factor of more than 4 over the last 2 years. Recall that Wellstock purchased $50 million of additional treasury stock in 2005; this purchase has the effect of decreasing the denominator of both the ROE and the debt/equity ratios, driving both ratios up. Is Wellstock now too debt leveraged? Probably not, as total debt (including short-term borrowing) is now only 21% of equity. Wellstock's strategy of using cash from operations to repurchase common shares looks like a sound one.

- This last conclusion is further bolstered by the ratio that appears just above the per-share data: capital expenditures seem quite steady at only about 22% of "net income plus depreciation and amortization"; that is operating cash flow before changes in working capital.

- Look now at the asset turnover ratios. Wellstock's is a capital-intensive business: total asset turnover is low and has not improved as the company's sales have grown. Fixed asset turnover (sales divided by fixed assets) show a modest improvement. We might usefully investigate further why Wellstock cannot achieve economies-of-scale in asset utilization as it is growing.

- What do the per-share ratios tell us? Wellstock's P/E ratio has improved (if we ascribe the 2001 P/E to abnormal circumstances). The current P/E—a bit above the average P/E for the New York Stock Exchange stocks in 2005, but not the high earnings multiple that might be accorded a high-growth company—suggests reasonably strong investor interest in Wellstock. Investors are not purchasing Wellstock's shares to achieve current return from dividends: yield is low and getting lower. During the last few years the median market price per share has grown faster than DPS. Has Wellstock's board decided to use excess cash to purchase treasury shares rather than boost the dividend (now 12 cents per quarter)? Given the steady decline in the payout ratio, probably so.

- Perhaps the most pronounced trend in all of these data is the last: total market capitalization divided by book value of equity; that is, market value per share times the number of outstanding shares divided by the book value of shareholders' equity—which is, remember, total assets less total liabilities! That ratio has increased by more than 30% per year. In this trend, we see the combined effect of (a) earnings improvements, (b) purchase of treasury stock (thus reducing equity book value), and (c) increased investor enthusiasm for Wellstock's shares (increased P/E).

- With a very high ROE and a low dividend payout ratio, Wellstock's sustainable growth rate is high, much higher than the growth rate that the company has accomplished over the past 5 years.

NEW TERMS

Accounts payable payment period	Ratios
Accounts receivable collection period	Return on assets (ROA)
Current ratio	Return on equity (ROE)
Dividends per share (DPS)	Return on sales (ROS)
Interest coverage	Times interest earned
Inventory flow period	Total asset turnover
Inventory turnover	Total debt to total assets
Long-term debt to total capitalization	Yield
Quick ratio	

CHAPTER 8

Cost Accounting
Digging Deeper into Valuation

You can think of **cost accounting** as a continuation of the discussion of valuation in Chapter 3. Our objective now is to value discrete activities within the organization, for example,

- manufacture 10 pieces of part number 4861,

- assemble unit 83 for Customer B,

- complete the design and documentation of new product T,

- complete the ordered repairs on the Saunders' 2003 Volvo station wagon,

- draw the legal documents required for Client Q to file suit against Company R.

The first two "jobs" occur in a manufacturing environment, the third in an engineering office, the fourth in an auto repair shop, and the fifth in a law firm. These five examples should help convince you that cost accounting is ubiquitous and you need to be generally familiar with its processes.

Manufacturing companies, from which the first two examples are drawn, convert materials from one "state" to another. These materials gain in value as they move toward product completion and ultimate shipment to the customer. That is, conversion adds value and the accounting system needs to track that increased value in order to properly value inventory and, eventually, cost of goods sold. Properly designed, the cost accounting system should also provide information useful to the manufacturing managers in evaluating operational efficiency.

Merchandising operations, from which the early chapters of this book derived most examples (recall Federated Department Stores), do not convert materials: Federated sells merchandise in the same state that it acquired the merchandise.

The third example above—engineering design and documentation—may occur within a manufacturing company or a product design or architecture or similar firm. The final design may be used internally or it may be sold. If sold, the accumulated cost is the debit to Cost of Goods Sold when the final design is delivered to the customer. Whether used internally or sold, the design cost information helps the engineering managers assess staff efficiency as well as the accuracy of their budgeting for this particular design task.

In the last two examples—auto repair shop and law firm—the cost accounting information typically is the basis for billing the customer, the Saunders' family for auto repair and Client Q for legal work.

COST ACCOUNTING IN MANUFACTURING

The typical manufacturing conversion process involves employees bringing together various materials, altering some of those materials, and assembling them into a final product that is shipped to the customer. Think of manufacturers of automobiles, consumer electronics instruments, furniture, pots and pans, microwaves, and the list is endless. The focus of some manufacturers is the processing of materials from one state to another: refining petroleum, producing paper or glass, making wine or beer, and again the list is endless.

To track the costs for these manufacturing activities, the cost accounting system must be able to identify and value the materials—**direct materials**—used as well as the hours of employee time—**direct labor**—devoted to the activities. Conceptually, this tracking is straightforward: the workers keep track of the time they spend on various tasks or jobs, and the material is counted, weighed, or otherwise measured. The execution of this simple concept can, however, be quite challenging: a great deal of data must be "captured" accurately and in a timely manner. Thank goodness for modern information technology!

Acquiring and processing the data on direct material and direct labor are necessary but not sufficient. Much more must be included:

- the cost of power to run the machines and heat and light the manufacturing facility,
- the salaries of supervisors who oversee not a single job but many simultaneously,
- the depreciation of the manufacturing equipment and facilities,
- the salaries of the maintenance and janitorial crews,
- IT expenses—equipment, staff, and materials that are ubiquitous on modern manufacturing floors,
- the salaries of the production and inventory control personnel,
- the cost of comprehensive insurance coverage.

How does the cost accounting system determine how much of these seven costs (a complete list would, of course, be much longer) should be assigned to a particular job or process? Well, if the supervisors are overseeing only one complex process, that is easy; but if the supervisors are overseeing tens or hundreds of jobs simultaneously, the record-keeping could be hopelessly complex. Similarly, in most manufacturing operations it is impractical or impossible to meter the amount of power, IT, or production control salaries utilized by each job.

You undoubtedly recognize these seven costs as examples of **overhead**. The **direct costs** discussed above can be identified directly with individual jobs or tasks. Overhead costs are, in contrast, often called **indirect costs**. Bear in mind that not all labor costs—i.e., wages, salaries, and related fringe benefits paid to employees—are direct costs; compensation paid to supervisors, janitors, production schedulers, IT troubleshooters, and so forth are included in indirect costs.

ACCOUNTING FOR OVERHEAD

If we cannot (or find it impractical or too expensive to) track these overhead costs to individual jobs or tasks, how do we deal with them? One possibility would be simply to give up the idea of tracking them in such detail: instead of matching them to the activity and thus including them in cost of goods sold, match them to the accounting period—just as are nonmanufacturing expenses such as selling and administrative expenses. That is, treat them as **period costs** rather than **product costs**.

This approach has the great appeal of simplicity! Its drawbacks are several:

a) The aggregate of these costs is large—and getting larger. One hundred or more years ago during the early part of the industrial revolution when cost accounting techniques were first developed, overhead costs were minor: simple machinery, limited supervision, no IT to speak of. Today, however, overhead costs typically overshadow direct costs; indeed generally *indirect labor* costs are higher than *direct labor* costs. And, this trend toward higher indirect costs and lower direct costs continues as more manufacturing activities are automated.

b) The accepted accounting principles discussed in Chapter 10 require that overhead be included in the valuation of inventories (in-process and finished goods inventories) and cost of goods sold.

Thus, these indirect costs (overhead costs) must be allocated to jobs in some rational—but necessarily arbitrary—manner. This allocation is accomplished by the use of an **overhead rate** applied to an **overhead vehicle**. We combine all of the indirect costs (elements of overhead) in a single bucket and spread them—like peanut butter—across all the activities (jobs).

But this process is complicated by the fact that inventory valuations and cost of goods sold valuations must be determined in "real time"—as the manufacturing occurs—and not simply at the end of the period when all overhead costs can be totaled up. Accordingly, estimates are required.

EXAMPLE: DJM MANUFACTURING

DJM manufactures safety helmets for use on construction sites. Its product line includes a host of styles and sizes. Its manufacturing jobs are defined in terms of a specified quantity of helmets of particular style and size. For simplicity sake, assume that its manufacturing overhead consists solely of the seven activities listed above.

As DJM approaches a new fiscal year, say 2007, its executive group forecasts sales for the coming year and translates that forecast into a production plan for the year: the number of helmets to be produced according to style and size. This production plan serves as the basis for estimating (budgeting) the amount that DJM will need to spend on each of the seven overhead elements. DJM's cost accounting staff now has the basis for determining its overhead rate for the coming year. DJM decides to assign the same amount of overhead to each helmet produced—spread the peanut butter evenly across the helmets! Thus, the number of helmets is DJM's overhead vehicle. By the way, it might have used hours or dollars of direct labor or some measure of direct material utilization as the overhead vehicle. Obviously, the vehicle must be common to all manufacturing jobs.

Here is how DJM derived its 2007 overhead rate:

2007 Overhead Budget ($000)

Power costs	$33
Supervisor salaries and benefits	216
Depreciation of manufacturing equipment	72
Maintenance and janitorial salaries and benefits	113
IT costs	69
Production/Inventory control salaries and benefits	82
Insurance premiums	12
Total overhead budget for 2007	$597

Planned production for 2007: number of helmets (000)   136

Overhead rate = 597 divided by 136 = $4.39 per helmet

In the common parlance of cost accounting, each helmet produced is said to **absorb** $4.39 of overhead. This process of *assigning* costs should not be assumed to imply that each helmet *caused* $4.39 of overhead. More on this point later in this chapter.

Note that included in this overhead rate are only costs/expenses for manufacturing; obviously, power and IT capabilities are also used by the sales, accounting, and engineering departments. Only a portion of DJM's total insurance premiums is appropriately assigned to manufacturing; the rest is included as a period expense and so, for example, is depreciation expense related to nonmanufacturing fixed assets.

Why not include these period expenses—selling, engineering, administrative (referred to in Chapter 2 as operating expenses)—and *absorb* those into the product as well? The objective of cost accounting is to determine what it costs to manufacture safety helmets, not what it costs to design and sell them and administer the overall organization. But, in fact, certain government contractors are required by their government customers to have two "overhead rates"—one to *absorb* the manufacturing overhead and the other to *absorb* operating expenses. With that exception, the rules of accounting require that expenses of manufacturing be separated (and included in cost of goods sold) from nonmanufacturing expenses (included below the gross margin line on the income statement).

You and I both know that DJM is unlikely to expend on these overhead expenses exactly what the company budgeted for the year. Yet the overhead rate is derived from the budget, not from actual expenses, in order that it can be used throughout the year to value inventory and cost of goods sold, month by month. DJM cost accountants do *not* go back at the end of the year and revalue inventory and cost of goods sold using a recalculated overhead rate. Instead, any difference between actual and budgeted overhead expenditures is identified as a **variance** and that variance adds to period expenses—if actual overhead expenditures exceeded the amount absorbed—or reduce period expenses—if expenditures were less than the amount absorbed. In any case, assuming that DJM folks are experienced at budgeting, these variances are likely to be small.

How often should overhead rates be recalculated? That depends on how volatile the business is. If DJM is long in the business of making safety helmets, and sales forecasts are generally reliable, annual recalculation of its overhead rate should be sufficient. Another company making "trendy" portable music devices may need to reset the overhead rate frequently, as its manufacturing processes improve and sales forecasts fluctuate.

How does DJM determine the cost of an individual helmet? Assume that the helmet was produced on Job 4762, which called for the production of 200 helmets of medium size and style V. All direct labor and direct material charges are collected for Job 4762; overhead is added (the overhead rate of $4.39 per helmet times 200 helmets). To determine the per-helmet cost, the resulting total cost of completing Job 4762 is divided by 200. This cost is used to increase the value of finished good inventory as Job 4762 is completed and the completed helmets move into inventory, and again to value cost of goods sold as these helmets are shipped to customers.

EXAMPLE: WARE AND FOSTER, A LAW FIRM

Ware and Foster (WF), a mid-sized law firm located in a mid-sized Midwestern city, provides a full range of legal services to its clients. WF uses a cost accounting system to determine the amount to be billed to individual clients. The lawyers themselves, and perhaps paralegals, constitute the direct labor of the law firm. The overhead vehicle is lawyer hours. Overhead consists of rent on the office, power and telephone expenses, depreciation on office equipment, insurance premium expenses, and so forth. Indirect labor included in overhead consists of wages and benefits paid to secretarial and other support staff, some portion of the managing partners' salaries, and probably the cost of legal continuing education. At this service enterprise, an allowance for WF profit may also be included in the overhead. The denominator of the overhead rate is the number of expected billable hours.

A potential client of WF might be told that he or she will be billed at the rate of $375 per hour of lawyer time used. More senior partners have a higher rate; associates (nonpartners) a lower rate. The lawyer may be well paid, but she does not receive the full $375 per hour. Overhead costs are not trivial in law firms.

Firms of accountants, architectural firms, plumbers, auto repair shops, and a host of other kinds of service enterprises follow cost accounting procedures analogous to those at WF. Plumbers and auto mechanics—as well as lawyers—take much abuse for their high hourly rates, but bear in mind that these rates include overhead expenses and profit.

UNDERSTANDING AND INTERPRETING OVERHEAD

These techniques for absorbing overhead are tidy, quite easily understood, and widely utilized. But the resulting information is too frequently misinterpreted, leading to incorrect—indeed, sometimes downright foolish—decisions. Here are several brief illustrations:

1. Suppose DJM anticipates that in 2008 a construction slowdown will occur, causing a corresponding reduction in safety helmet sales. When in late 2007 DJM reviews its 2008 budget of overhead expenses, it estimates that, despite a 15% decline in the sales forecast, the company can realistically reduce its manufacturing overhead budget by only about 5%. Accordingly, its overhead rate will increase from $4.39 per helmet in 2007 to $4.89 in 2008, a 50-cent increase per helmet. The manufacturing vice president urges that DJM increase its safety helmet prices by an amount sufficient to cover this added overhead per unit.

2. In 2007 a design engineer recommends that one step of the assembly process be subcontracted to a firm in Mexico to take advantage of that country's much lower labor rates. He argues that DJM's labor costs per helmet would be decreased by 87 cents and its manufacturing overhead would be reduced as well. The Mexican subcontractor offers a price of 99 cents per part (one part per helmet). The manufacturing vice president strongly objects, noting that the firm's overhead rate would continue at $4.39 per helmet and its direct costs (labor and material) would increase by 12 cents.

3. In 2007 a Brazilian customer contacts DLJ with a request that DJM alter its design of a particular safety helmet and custom manufacture 25,000 helmets of this altered design. DJM currently has no sales in Brazil and anticipates none in the next few years. The Brazilian customer seeks a low price, arguing that this added volume will absorb a great deal of DJM's manufacturing overhead. DJM has the capacity to produce this large order (equal to nearly 20% of planned 2007 production) with the use of some overtime. Wide disagreement breaks out among DJM executives as to whether this order should be accepted.

4. In late 2009 DJM anticipates a recovery in construction activity in 2010 and adjusts its sales forecast and overhead rate accordingly. In fact, the recession continues throughout most of 2010 and DJM several times during the year has to reduce its production plan to better match the depressed sales volume. At the end of that year DJM has a "negative" overhead variance (that is, actual expenditures on overhead exceeded by a considerable amount the overhead absorbed by the safety helmets). DJM's president criticizes the vice president of manufacturing for his poor control of overhead costs.

These four scenarios are discussed in the following four sections. By the way, parallel scenarios could be constructed for service businesses.

Pricing Based on Cost Accounting Information

Despite economists' consistent advice to the contrary, managers seem persistently tempted to "price" based upon "cost": take the cost as determined by the cost accounting system and add a standard mark-up percentage. But customers have no interest in the manufacturer's cost; they simply compare prices among competitors.

Note that the pricing rule DJM's vice president of manufacturing is recommending in scenario 1 above will cause prices to increase when sales are weak, thus causing further sales weakening, and decrease when sales are robust and further stimulation of sales may be unnecessary or even unwise. In fact, DJM should consider the exact reverse.

Why might DJM expect overhead costs to decrease only 5% when volume of manufacturing decreases by 15%? Because many of the overhead expenses will not change appreciably, if at all, with a 15% decrease in manufacturing activity. (By the way, DJM could probably increase production by 15% and incur only a modest increase in overhead.) Most of these expenses are, in that sense, **fixed**. For example, depreciation expenses, insurance premiums, and maintenance expenses will probably be unaffected by modest changes in safety helmet output. Perhaps, on the other hand, the first two—power and supervision—will increase or decrease as volume of activity grows or shrinks; if so, power and supervision are **variable expenses**. (This clean, bright-line separation of fixed and variable expenses is, in fact, stylized and unrealistic—but useful for illustrative purposes; many of these expenses are likely to be semivariable.) These simplified assumptions lead to the conclusion that about half of the budgeted overhead dollars is variable and the other half is fixed. In most manufacturing operations, fixed costs are an even higher percentage of the total, and the percentage increases as more activities are automated.

As we consider the other three scenarios, we will see that misunderstanding of the behavior of overhead costs—some fixed, others variable—with changes in volume is at the root of the poor recommendations being made. But before we move on, let me reemphasize the definition of expense variability here. Expenses may change for a whole host of reasons, generally driven by specific management decisions. Here we are talking about the extent to which expenses will perforce change (vary) or remain the same (fixed) with modest changes in volume of activity.

Analyzing Subcontracting Opportunities

The safety helmet design engineer is right: some of DJM's overhead will be reduced if it subcontracts one step in the manufacturing process. We identified power and supervision expenses as variable. The correct decision regarding subcontracting turns, then, on whether DJM can anticipate at least a 12-cent per helmet reduction in overhead.

Misinterpretation is more common if direct labor is used as the overhead vehicle. Suppose that DJM's vehicle was "dollars of direct labor" rather than "safety helmet units" and that its overhead rate was set at $2.24. Now the temptation is to conclude that for every dollar of direct labor saved—by improved efficiency, product redesign, or subcontracting—DJM saves $1.00 in direct labor plus $2.24 in overhead, a total of $3.24. Not so. DJM's variable manufacturing overhead will be reduced slightly, but its fixed overhead not at all. Assuming here that half the total overhead budget is variable, we could anticipate that, if subcontracting saves 50 cents of direct labor, it should save $(0.5 \times 0.5 \times 2.24 =)$ 56 cents of overhead. Based on these assumptions, subcontracting seems to make economic sense, assuming that quality, delivery, and other aspects of the subcontracting arrangement are satisfactory.

Full Costs Are Not Incremental Costs

An important point is, then, that "costs" derived from typical cost accounting systems are not incremental costs; that is, they do not tell us how much more expense DJM will incur in building one more helmet, or save by building one less helmet.

The third scenario calls for that information: the incremental cost. If our assumptions as to fixed/variable cost behavior are reasonably correct, the incremental cost is approximately direct labor plus overtime premium on wages plus direct material plus variable overhead, where per-unit variable overhead is one-half of $4.39, or $2.20.

So, that cost figure—incremental cost—is the place to start the analysis of whether to accept the large Brazilian order for 25,000 helmets. If DJM will incur no other additional costs (how about the cost of redesign, shipping finished product to Brazil, different packaging, and so forth?), then, if the price offered by the Brazilian customer exceeds the incremental (variable) cost of producing the helmets, DJM can conclude tentatively that its total profitability will be better if it accepts this order than if it declines the order.

This conclusion should be tentative until it considers a number of related issues. First, while DJM may be willing to offer some discount, it will not want to discount its regular prices all the way down to incremental cost; it wants to make a reasonable contribution to profit. Second, might the acceptance of this low-priced order annoy some of DJM's other customers, including its distributors? Third, might some of these helmets shipped to Brazil find their way back into the U.S. market, thus displacing full-margin sales that DJM might otherwise have enjoyed? Fourth, might a sales surge in the coming year push production activity to full plant capacity? If so, DJM would not want to run the risk of having to turn away full-margin sales because its manufacturing facility is so occupied with this low-margin Brazilian order.

This scenario points out that incremental costs are typically more useful to managers as they make day-to-day operating decisions than are so-called **full-absorption** costs: costs that include an allocation of fixed overhead as well as of variable overhead expenses. Could we design an accounting system that routinely delivered this information? Yes; such systems are called **variable costing** systems, and they are increasing in popularity in manufacturing environments. These systems treat fixed overhead expenses as period expenses and define product expenses as including only direct material, direct labor, and an appropriate allocation of *variable* overhead expenses.

Interpreting Overhead Variances

Variable costing systems have the added advantage of simplifying the interpretation of overhead variances. Any variance is due solely to actual expenditures on (variable) overhead elements being greater or less than the amount budgeted. This variance information is useful to the manufacturing managers.

Unfortunately, full absorption of overhead complicates the task of interpreting overhead variances. Describing the complication is itself quite complicated!

Suppose for the moment that all overhead in a particular operation—for example, a law firm—is fixed; small variations in the amount of legal work billed to clients in a particular accounting period has no effect on what the law firm needs to spend on its various overhead elements. By the way, that description of overhead cost behavior in law firms is probably pretty accurate. The law firm's budget for 2006 was as follows:

Budgeted hours of billed time	16,000
Budgeted overhead expenditures	$2,400,000
Therefore, the overhead rate for 2006	$150 per billed hour
Actual results for 2006 turned out to be:	
Overhead variance	$10,000 credit (favorable)
Actual hours of billed time	16,100

In cost accounting parlance, the law firm has **overabsorbed** overhead by $10,000.

The firm's managing partner congratulates herself and her colleagues for good control of overhead spending—witness the $10,000 credit overhead variance.

But the overhead variance combines two effects: (a) spending control and (b) volume, that is the number of billed hours. The law firm *absorbed* (16,100 × $150 =) $2,415,000 of overhead, $15,000 more than anticipated solely because it billed an extra 100 hours of lawyer time. Therefore, if the spending on overhead (remember, all overhead elements are *fixed*) had been on budget, the total overhead variance should have shown a $15,000 credit, or overabsorbed, variance. In fact, the firm's overhead expenses were ($15,000 less $10,000 =) $5,000 over budget. While this is a very minor budget overrun (well under 1% of the budget), the managing partner clearly does not understand how to analyze the overhead variance.

When overhead expenses for an operation consist of both variable and fixed elements—a typical condition—the analysis of overhead variances is somewhat more complicated. When operating activity exceeds budget, the variable expenses should be expected to exceed the original budget; we need to volume-adjust the overhead expense budget before we can separate the (a) spending and (b) volume effects, which are rolled up together in the overhead variance account.

In short, whenever an overhead rate is used to *absorb* or assign costs that include fixed expenses—that is, expenses that do not change with modest changes in volume of activity—extra care must be taken to separate spending and volume effects when interpreting overhead over- or underabsorption.

STANDARD COSTING

Another widely used embellishment on plain-vanilla cost accounting systems, one that supports "management by exception," involves the use of **standard costs**. Consider a manufacturing company that produces standard products in high volumes, say, laptop computers. That company may decide to use a standard cost accounting system so as to enhance its ability to detect when and where its operations are deviating from plan. To do so, it requires that the manufacturer establish "standards costs" for its products—estimate the quantity and cost of each material component and estimate the number of hours and wage rates for all direct-labor employees who will fabricate parts and assemble them into the finished laptop computer. Then, as the computers are produced, the manufacturer can compare—that is, isolate in variance accounts— the difference between the standard and actual costs. Variances can be constructed to highlight those differences that management wishes to monitor particularly diligently. For example, managers may decide that scrap has the potential to be a significant problem but variations in price of purchased components (because of long-term supply contracts) do not. The standard cost accounting can be designed to highlight variances in material usage, but ignore differences between standard and actual prices paid for components.

If standard costing sounds complicated and expensive, it typically is not—if the manufacturer is producing standard products in large quantities. The first advantage is that inventory and cost of goods sold of the laptop computers are valued at standard cost, not actual cost; different actual costs do not have to be tracked through inventory and cost of good sold. The variances between actual and standard costs/prices are drawn off as production progresses. Second, the "standards" themselves serve useful functions in addition to their use in cost accounting; for example, standard labor hours assist production scheduling and standard material prices are useful guides to the purchasing agents. And, finally, the variance information itself can be enormously important to manufacturing managers as they monitor the processes in their factories. Standard products produced and sold in high volumes typically encounter severe price competition in the marketplace; information that can assist managers in improving manufacturing efficiency is likely to be very useful.

Here is a quick example of labor variances that can be generated by a standard cost accounting system; similar variances could be generated for material costs. Suppose that DJM determines that the standard labor hours to produce a particular helmet is 0.304 and DJM's standard labor wage rate is $14.36 per hour, including fringe benefits. Obviously, the standard labor cost per unit is (0.304 × $14.36 =) $4.365. Now suppose that 100 of these units are produced on Job 5876 and the total labor charged to Job 5876 was 31.5 hours and $432. The total labor cost variance for this job is then

Actual labor cost	$432.00
Standard labor cost	(100 × $4.365 =) $436.50
Total labor variance	$4.50 credit.

This single labor variance can then be separated into its "wage rate" and "labor efficiency" components as follows:

a) Actual labor cost (actual hours × actual wage rate) $432.00

   Actual labor hours at standard wage rates .... (31.5 × $14.36 =) $452.34

   Labor wage rate variance .... $20.34 credit.

b) Actual labor at standard wage rates $452.34

   Standard labor at standard labor wage rates .... (100 × $4.365 =) $436.50

   Labor efficiency variance .... $15.84 debit.

The sum of the two variances, one debit and one credit, is, of course, $4.50 credit, the amount of the total labor variance calculated earlier. It is useful to DJM's manufacturing managers to know that on this job more hours were spent than standard (plan), but this was more than offset by the wage rates paid being below standard (plan).

A variance can be calculated for every situation where a standard, or budget, is set. Consider another example. Suppose that DJM provides its field sales force with extensive pricing discretion, and the sales manager wants to track (by salesperson and in total) the pricing discounts agreed to. Assume that DJM sets its standard price for a certain helmet model at $57.10 and that actual sales for a particular accounting period were 1,320 helmets and $73,860. The sales price variance—the total price discounts allowed—is then

Actual sales at standard price   (1,320 × $57.10 =) $75,372

Actual sales at actual prices $73,860

Sales price variance $1,512 debit.

We are tempted to label this (and other) debit variance as "unfavorable" and to label credit variances as "favorable". Be careful. For example, one of DJM's field salespersons might generate a "favorable" sales price variance, but if in so doing he lost many orders to the competition, the overall result may indeed not be "favorable." Variances simply point out where actual results differed from plan, and assigning value judgments to resulting debit and credit variances can be misleading.

SUMMARY

Cost accounting techniques are usefully applied to many activities. Determining the cost of products as they are "converted" to their finished states is essential in manufacturing enterprises in order to value inventory and cost of goods sold. Building contractors, law firms, accounting firms, consultants, research institutes, and others use these techniques as the basis for billing clients. But cost accounting's most important benefit is in generating information useful to management in monitoring and controlling costs.

Manufacturing managers need information to allow them to assess the efficiency of their operations—the amount of direct labor used vis-à-vis the plan; the value of material used versus plan; the control of overhead expenses. Engineering managers want feedback on hours spent by individual projects. In any environment where the jobs, tasks, or projects are clearly definable and budgets (hours and dollars) are established in advance, cost accounting provides feedback that allows comparisons of "budget versus actual." In turn that feedback permits "management by exception": management can focus attention on those activities that are not proceeding according to plan.

NEW TERMS

Absorb (absorption)

Cost accounting

Direct cost

Direct labor

Direct material

Fixed expenses

Full-absorption costing

Indirect cost

Over- (under-) absorbed overhead

Overhead

Overhead rate

Overhead vehicle

Period costs

Product costs

Standard costs (costing)

Variable expenses

Variable costing

Variance

CHAPTER 9

Budgeting and Forecasting
The Past Is Prologue

A major reason to better understand the financial score is to gain sufficient knowledge about an organization's past so as to project its financial future with more accuracy and reliability. This chapter presents some guidelines—indeed some are almost clichés—about budgeting and forecasting. With experience in analyzing financial statements comes a growing instinct to extend present trends in financial performance and position into the future; to support that instinct, most annual reports present detailed financials for the past several years and many provide financial highlights for the past 10 years. Recall the discussion of trends in financial ratios in Chapter 7. And recall from Chapter 8 that determining an overhead rate requires both a budget of the overhead cost elements for the coming period and a forecast of business activity—building products or performing services—for that period.

OPERATING BUDGET

Every well-run organization develops **operating budgets**. Even if the budget is not committed to paper—a foolish oversight—a budget is implicit. Sometimes this implicit budget is simply that next year's performance should track this year's. Good managers, however, insist on taking the opportunity, at least once per year, to consider how operations in the coming period might be improved: raise revenue, reshape expenses, drop an obsolescing and high-cost product, invest in new development or marketing or IT initiatives, generate more cash and pay down debt, and many other possibilities.

The frequency of rebudgeting depends on the volatility and predictability of the organization's activities. A new internet company or consumer electronics manufacturer will need to budget very frequently, probably at least quarterly. An electric power utility of manufacturer of branded, widely accepted food products will probably find that a redo of the budget annually is sufficient.

Why bother? Why do we not just count on everyone doing his or her best? Most people dislike the budgeting process; it is difficult, often contentious, and necessitates personal commitments (to achieve certain sales or meet certain expense targets) that can be uncomfortable. But three good reasons for budgeting operations are as follows:

1. Plans—and, yes, commitments—become explicit. Platitudes, arm-waving and optimistic cheerleading are at least tempered and often contradicted when plans are committed to paper and dollars and cents consequences are attached to them.

2. Communication throughout an organization is facilitated by explicit budgets by sales region (or even by individual salesperson), expense center, division, and so forth. Segments of the business do not operate in isolation; they depend on—and are in turn depended upon by—other segments who need to know their plans.

3. Budgets become benchmarks against which actual results are compared. Managers can then focus their attention on variances—in either direction—between budgeted and actual results, thus facilitating so-called **management by exception**.

In connection with the third advantage, it is worth reminding ourselves that an "exception" is not necessarily "bad," even if it reduces net income. Every exception deserves careful analysis, even when it results in *increased* net income, because it typically leads to new knowledge about the organization itself, its customers, and/or its competitors. The temptation to "penalize" an individual or group for an exception that reduces profits should be resisted for three primary reasons. First, the causes of the exception may be entirely beyond their control. Second, the exception may in fact be in the long-term (but not necessarily short-term) best interests of the company. And third, penalties or the threat of penalties can lead to some very dysfunctional and even unethical employee behavior: cover-ups, finger-pointing, unrealistic future budgets, short-term expedients that have excessive long-term costs, and so forth.

Budgeting of operations generally commences with a thoughtful estimate of revenue for the coming period, say a year. These revenue estimates are best built up from the bottom—**bottom-up budgeting**: the preliminary estimate is the compilation of what each salesperson and then each sales region (or different service function) will achieve over the course of the year. These individual estimates or plans are "negotiated" as they move up the chain of command in the organization—those that are too timid or cautious are boosted and those that are too aggressive or optimistic are lowered.

But aggregated revenue estimates need to be modified or adjusted for conditions or plans that the field sales troops are unaware of:

- new product introductions or delays in introduction,
- shifts in marketing emphasis or the initiation or cessation of marketing initiatives,
- anticipated pricing changes,

- macroeconomic forecasts of industry conditions,
- expected competitor actions or reactions.

The revenue forecast is the cornerstone of the operating budget. The organization exists to serve its "customers" (sometimes called students, clients, parishioners, audiences). Nonsales functions must be organized, staffed, planned, and optimized—that is, budgeted—to provide those services and/or products. This budgeting is also best done bottom-up.

Of course, sometimes a mismatch develops between revenue expectations and planned expenses. Negotiations resume, involving more people and often more tension. If required customer service cannot be delivered at affordable expenses, something has to give: perhaps it is the revenue expectation, perhaps it is expense and staffing levels, and often it is some of each. Finally, a budget is hammered out that gains the approval of senior management and the governing board.

You surely realize that when individual performance bonuses are geared to "achieving budget," budgeting becomes more complicated, more negotiations occur, and people may be motivated toward both functional and dysfunctional behavior.

PRO FORMA FINANCIAL STATEMENTS

The completed and approved operating budget typically is—or can easily be—translated into a **pro forma** income **statement**. That is, the operating budget projects the income statement for the coming year. Note that a pro forma income statement requires estimates beyond just revenues and expenses. The applicable income tax rate must be forecast. The granting and exercising of employee stock options must be estimated (as well as the company's stock's market price, which will influence the earnings-per-share dilution calculation). Noncash expenses and employee long-term benefits must be estimated. Changing interest rates will affect both interest income and interest expenses.

Comparing this pro forma income statement to the enterprise's recent statements of operations, that is, its historic income statements, can test its reasonableness. As implied in Chapter 7, comparing the percentage that each expense category is of revenue is revealing. If the pro forma statement forecasts a sharp increase in gross margin percentage, management should be able to articulate the reasons why this improvement is achievable: for example, price increases, increases in volume of activity with proportionally lower increases in fixed expenses, new processes perhaps as a result of automation, or a shift in product mix in favor of more profitable products.

Similarly, abrupt improvements in marketing or administrative expenses as a percent of revenues are either unrealistic or they are explainable.

With a good pro forma income statement in hand, a forecaster can turn attention to a pro forma balance sheet. Here, the ratios discussed in Chapter 7 are very handy, as are the company's past trends in these ratio values. For example, if a company forecasts a 20% increase in revenue selling the same set of products or services to the same set of customers, it should anticipate a 20% increase in investment in accounts receivable unless it forecasts a change in the accounts receivable collection period. Of course, if the company expects to sell to less creditworthy customers, or offer its customers more generous terms of sale or invest in more aggressive collection efforts, these changes are reflected in an altered assumption as to accounts receivable collection period.

Ratios are also distinctly useful in estimating inventory (inventory turnover ratio), accounts payable (A/P payment period), and accrued liabilities; that is, elements of working capital. Forecasts of long-term assets and liabilities require a thorough understanding of just what investments in noncurrent assets are anticipated and what financing transactions (new borrowing, debt repayment, sales of equity, purchase of treasury stock, dividend payments) are planned.

With a complete pro forma income statement and balance sheet in hand, constructing a pro forma cash flow statement is a straightforward, but useful, exercise. Running out of cash is a disastrous event for any organization—or indeed for you or me! We will look further at cash budgeting later in this chapter.

Once again, we need to be alert to possible mismatches or inconsistencies among the three pro forma statements: income, balance sheet, and cash flow statement. Perhaps the company will be unable to obtain the additional resources required to invest in fixed assets (or working capital) that would be necessitated to achieve the revenue projections shown on the pro forma income statement. Or strong cash generation in the coming year may permit a major reduction in debt, and thus lower interest expenses than projected on the pro forma income statement. Or the increased financial leverage evidenced in the pro forma balance sheet is deemed by the board of directors to be simply too risky. Or heavy investments in fixed assets coupled with high dividends expected by the shareholders will result in an unacceptably low end-of-year cash balance.

EXAMPLE: WELLSTOCK CORPORATION

Picking up the example we looked at in Chapter 7, Wellstock Corporation, we can use the assumed forecasting data listed in Exhibit 9-1 to develop pro forma financial statements. Wellstock's internal company forecasters would, of course, have access to a great deal more information on many more parameters—to say nothing of a comprehensive "bottoms-up" sales forecast—upon which to base their estimates. Thus, what follows is simply an abbreviated example of budgeting and forecasting.

Exhibit 9.1: *Wellstock Corporation Forecasting Parameters, Fiscal Year Ending September 30, 2006*

Sales growth

  Products   23%

  Services   14%

Cost of Sales

  Products   72.7% of revenue (compared to 73.0 in 2005)

  Services   71.1% of revenue (same as 2005)

Operating expenses

  Selling, general & administrative expenses: 11% of sales plus

  $4 million for the new marketing initiative

Interest expense   to be determined, depending on borrowing required

Income tax expense   31% of pretax profit

Investment in fixed assets $15.5 million

Dividend payout rate   25% of net income

Shares to be repurchased   1.75 million

Average (market) price of repurchases $46.50 per share

Forecasting a strong general economy and a continued increase in market share, Wellstock is budgeting another significant increase in both revenue and profit. Top management believes that 2006 is the logical year to invest aggressively in both marketing and automation (fixed assets). On the other hand, because so many of its manufacturing and operating expenses are fixed rather than variable, the company expects that it can reduce the percentage that these expenses represent of revenue (before considering the new marketing initiative). Wellstock does not expect to have to recognize any goodwill impairment in the coming year. It expects that the company's income tax rate will return to historic levels or higher. In recent years Wellstock's yield (dividends as a percentage of market price) has declined; management feels that this situation has retarded appreciation in market price and believes that a 25% payout ratio (total dividends divided by net income) would be viewed very favorably by security analysts. Given the company's relatively low debt leverage and expectation of strong cash flow generation, the company plans to step up its share repurchase program (increase in treasury stock) by acquiring 10% of its outstanding shares; if necessary to accomplish this buyback, it is prepared to increase year-end total debt to 25% of equity. Before proposing to Wellstock's board of directors this stepped-up buyback, top management is eager to prepare and analyze the forecasted financial statements. The company does not currently anticipate making any major acquisitions in the coming year; however, should an attractive acquisition opportunity arise, the stock buyback program would be dialed back to free up the resources to effect the acquisition.

The primary elements of working capital should grow in line with the company's overall growth, as the company anticipates no change in the accounts receivable collection period, accounts payable payment period, and inventory turnover; employment will grow at about the rate of revenue growth (including new personnel in marketing) and thus accrued liabilities should maintain its historic relationship. Wellstock anticipates that, for reporting purposes, it will depreciate the fixed assets to be acquired this year on a straight-line basis over a 10-year life. About $4 million of the principal balance of the long-term debt is due in 2006 and the retirement benefits liability should grow by about $7 million. If indeed the market price of the company's common stock responds favorably to both corporate growth and increased yield, exercise of stock options by company employees is forecast to grow to about $15 million.

With these assumptions and forecasts of the parameters shown in Exhibit 9-1, Wellstock's corporate finance staff should have sufficient information to complete pro forma statements. Of course, this exercise will need to be iterative: for example, the final forecasted cash balance on the balance sheet, interest expense, and the number of shares outstanding at year-end will depend on how many shares the company decides it can afford to repurchase and what additional borrowing it decides to undertake to support the repurchase program—and of course those decisions will be driven by the results of the forecasts.

Exhibits 9-2 and 9-3 present the forecasted income statement for fiscal year 2006 and the balance sheet at year-end, September 30, 2006. The preliminary projections of revenue and cost of sales derive directly from the assumptions shown in Exhibit 9-1. However, the estimate of product cost of sales should probably be boosted by the expected depreciation on the new automated equipment to be acquired; assuming that on average this new equipment goes into service at mid-year, however, this amounts to only about $1 million (5% of $15.5 million cost). With a revenue projection of $419 million, the operating expenses (selling, general and administrative) can be forecast at [(11% × 419) plus $4] $50 million. The forecasters need to make a tentative assumption as to interest expense (say, $1 million increase) and interest income (no change) subject to revision once the balance sheet and cash flow statement reveal how much needs to be borrowed. Wellstock's income tax rate has hovered around 30% (of profit before tax) until 2005 when it dropped several points; with a nod to conservatism, the forecasters decide to boost the rate to 31%.

Estimating earnings per share and dividends per share is premature, as the forecasters need to go further before they can estimate the total number of shares outstanding. Moreover,

Exhibit 9.2: *Wellstock Corporation Income Statement Forecast, Fiscal Year 2006 ($ millions)*

Sales	
Product sales	$377
Service sales	42
Total sales	419
Costs, expenses and other	
Product cost of sales	275
Service cost of sales	30
Selling, general and administrative	50
Interest income and expense	—
Total costs, expenses and other	355
Income before income taxes	64
Income tax provision	20
Net income	$44
Earnings per share, basic	to be determined
Earnings per share, diluted	to be determined
Cash dividends per share	to be determined

these per-share estimates are not required to complete a preliminary pro forma balance sheet and cash flow statement.

The pro forma (budgeted) income statement for the next year, shown in Exhibit 9-2, viewed in its entirety, appears pretty unambitious: overall sales growth in excess of 20% but only a 10% increase in net income. After a review by top managers, the operating managers and budget folks are likely to be sent back to have another look: should the new marketing initiative be spread out over two years; can the automation equipment reduce product cost of sales by more than 3/10 of 1% in the coming year; perhaps the forecasters have (at this early stage) been too pessimistic about interest income and expense in the coming year, and so forth?

Nevertheless, let us accept this preliminary pro forma income statement to move on to forecast the balance sheet—see Exhibit 9-3.

Our first conundrum is that, until the pro forma cash flow statement is completed, the forecaster must assume a tentative value for cash and cash equivalents. Let us assume that Wellstock's treasurer feels that the minimum year-end cash balances should be, say, 2.5% of annual revenue; any amount below that level makes her and other executives nervous: normal

Exhibit 9.3: *Wellstock Corporation Balance Sheet Forecast, September 30, 2006 ($ millions)*

ASSETS

Current Assets	
Cash and cash equivalents	$ 10
Receivables	88
Inventories	84
Current deferred income taxes	24
Other current assets	4
Total current assets	210
Property	54
Intangible Assets	11
Goodwill	46
Other Assets	39
Total Assets	$360

LIABILITIES AND SHAREOWNERS' EQUITY

Current Liabilities	
Accounts payable	$35
Compensation and benefits	33
Income taxes payable	5
Product warranty costs	21
Other current liabilities	48
Total current liabilities	142
Long-Term Debt	16
Retirement Benefits	83
Other Liabilities	8
Shareowners' Equity	
Invested capital	141
Retained earnings	50
Treasury stock	(80)
Total shareowners' equity	111
Total liabilities and shareowners' equity	$360

cash flow volatility might cause embarrassing cash shortages. Think of this amount as a kind of "safety stock" of cash. If the final pro forma projections reveal the availability of more cash, that excess will be invested in cash-equivalent securities. Accounts receivable and inventory are projected simply by taking the actual 2005 collection period and inventory turnover and applying them, respectively, to 2006 projected sales and cost of sales. Assume that the deferred income tax asset increases by the same percentage that 2006 income tax expense increased from 2005. Finally, the amount of other current assets is now trivial and likely to remain so; to guard against overoptimism in projecting cash flow, perhaps we should increase this by $1 million.

The projection of the value of property requires us to estimate the depreciation expense for 2006 on currently-owned fixed assets (which lowers this value) and the new investment, assumed to be $15.5 million, less the $1 million of depreciation on these new assets to be recognized in 2006. We assumed earlier that Wellstock will not recognize any impairment on its current goodwill balance, nor will it make any acquisitions that might bring additional goodwill on to its balance sheet. Lacking any better information, let us assume that Wellstock's intangibles and other long-term assets stay the same; note that this is an aggressive (the opposite of conservative) estimate, as any increase will consume cash.

For simplicity sake—we would be a lot more knowledgeable if we were inside Wellstock—assume that all the current liabilities increase by the same percent as revenue increases (which is expected to be matched by the growth in personnel).

Now we come to the tough part: projecting permanent capital, that is, long-term debt and owners' equity. Iterating will be essential. For this early estimate, simply apply the assumptions made above: the $4 million principal repayment required in 2006 reduces the long-term debt and retirement benefits increase by $7 million. We again encounter an "other" category; we increased the "other assets" by a bit in order to be conservative (increases use cash), but now to be conservative we lean in the direction of lowering the amount (increases provide cash). Let us leave it the same as at year-end 2005, $7 million.

We are now down to owners' equity. Invested capital will increase only by the amount of the employee stock options exercised (assumed to be $15 million) and retained earnings will increase by 75% of projected net income, the other 25% being paid out in dividends. For the moment, let treasury stock be the "balancing item"—after all, a major question facing management is whether Wellstock can repurchase 10% of its outstanding shares. Obviously, total assets must equal total liabilities, and working from that total we can deduce that Wellstock can afford to have $80 million of treasury stock at year-end 2006, an increase of $31 million. Taking the assumption of $46.50 average price per share to be acquired, $31 million will permit Wellstock to acquire about 667,000 shares, a little less than 4% compared to the 10% of outstanding shares that was the company's original goal.

Pushing the analysis one step further, we can calculate that, at the assumed average share purchase price, Wellstock would need to borrow over $50 million in order to repurchase 10% of its outstanding shares. To stay within its guideline of maximum long-term debt at 25% of owners' equity, Wellstock could borrow only another [(0.25 × 111) − 16] $12 million. In turn, that borrowing would permit the company to repurchase about another 1.5%. Perhaps Wellstock should reduce its goal for the coming year to repurchasing 5% rather than 10% of its shares. But that additional borrowing is likely to boost interest expenses by over $1 million, and with $5 million less in cash-equivalent securities, interest income will be lower than that of last year, and both will negatively affect net profit. But, if 5% of shares are repurchased, both earnings per share and dividends per share will increase, perhaps driving up the shares' market price, thus requiring still more cash for the repurchase program. By now you have grasped why financial forecasts are iterated numerous times.

These forecasts also facilitate sensitivity analysis: testing the consequences should estimated parameters not be achieved. And, if the forecasts are computer based, analyzing "what if" questions becomes a simple and highly useful exercise. What if (a) sales do not grow as anticipated; (b) prevailing interest rates rise or fall; (c) key suppliers demand higher prices or more timely payment of invoices; (d) the average stock price moves to $55 per share, rather than the assumed price of $46.50 or, because of a slump in the overall market, falls to $32 and far fewer employee stock options are exercised; (e) Wellstock's prime competitor lowers prices by 10%; (f) Wellstock's labor union threatens a strike that might close down production for a month—the list of "what ifs" is virtually endless.

CASH BUDGETING

Once Wellstock's management and board make key decisions such as its 2006 dividends, its 2006 stock repurchase objective, its willingness to increase long-term debt, and so forth, we could develop a final set of pro forma statements, including a cash flow statement. The latter would show us the net increase or decrease in net cash as of the end of 2006 (September 30).

But another company—particularly one whose revenues and expenses are seasonal—would want to track cash balances throughout the year, not just at year-end, to be certain that its safety-stock cash balances never fall to a dangerous level or that it could borrow any shortfall under loan agreements now in place. This is best accomplished by a month-by-month (or conceivably over even shorter intervals) projection of cash inflows and outflows. Here is a simple and short example.

Exhibit 9-4 shows some assumptions for a 7-month period for a specialized beach-furniture store located in a popular beach town. Obviously its primary sales occur in the period June through September, with steady but modest $15,000 monthly sales over the other 8 months. The store wants to be well stocked before its busy season commences. It employees

Exhibit 9.4: *Seasonal Beach-furniture Store Cash Flow Projections, April through October ($ 000)*							
ASSUMPTIONS	**APR.**	**MAY**	**JUNE**	**JULY**	**AUG.**	**SEPT.**	**OCT.**
Expected sales	$15	$15	$20	$35	$40	$25	$15
Expected shipments rec'd	15	20	25	25	10	5	10
No. summer (temp.) empl.	–	–	2	3	3	1	–
Other Cash Outflows:							
Salaries, permanent ees.	3	3	3	3	3	3	3
Rent, utilities, etc.	2	2	2	2	2	2	2
MONTHLY CASH FLOWS							
Customer cash receipts	15	15	15	15	20	35	40
Payments to suppliers	15	15	20	25	25	10	5
Salaries to summer employees	–	–	2	3	3	1	–
Other cash outflows	5	5	5	5	5	5	5
Monthly cash flow	(5)	(5)	(12)	(18)	(13)	19	30
MONTH-END CASH BALANCES							
	–	(5)	(17)	(35)	(48)	(29)	1

college students at a salary of $1,000 per month to augment its store sales staff during the busy period. During the off-season months, the store carries a "normal" inventory. The store employs a few year-around personnel. With a gross margin of 33%, the store just breaks even during the off-season, but anticipates an $11,000 operating profit over the 4-month busy season.

To meet the terms of sale offered by competitors, this store provides customers extended, no-interest terms; on average, customers take 60 days to pay their bills. On the other hand, suppliers insist on payment in 30 days. Since the store plans to have a beginning cash balance of only $5,000 at April 1, it will need to borrow to cover its seasonal cash needs.

In November the store will have a positive cash flow of $15,000 and therefore a month-end cash balance of $16,000—equal to its $11,000 seasonal profit plus its original $5 cash balance.

These cash projections indicate that the store needs to be able to borrow $48,000 at the end of August, right at the end of its busy season, an amount that it should be able to fully repay within 60 days, the end of October. This store needs an accommodating bank! (Incidentally, interest on this debt is ignored in these cash flow projections.)

NEW TERMS

Bottom-up budgeting

Cash flow projection

Management by exception

Operating budget

Pro forma statement

CHAPTER 10

Rules and Integrity
Are the Books Cooked?

The past decade has witnessed unprecedented financial misbehavior on the part of financial officers, chief executives, and auditors. Therefore, this final chapter in a book dedicated to helping you better understand financial statements appropriately addresses the issues of institutional integrity and financial veracity.

In this chapter, we will explore first the influences that the ethical standards and culture of a corporation have on its financial statements. Then we will review both externally mandated accounting rules and the role that independent auditors play in assuring fair presentation of a corporation's financial position and performance. We turn next to a brief review of the regulators focusing on protecting shareholders and investors from unscrupulous corporate behavior, and conclude with a brief review of how "books" can be "cooked"—and, of course, a strong urging not to "cook" them!

ETHICS, VALUES, AND CULTURE

Many of the earlier chapters—particularly Chapters 3 and 4 addressed to valuation and timing issues—emphasize that managements and accountants have broad discretion in valuing assets, liabilities, revenues, and expenses. Within this broad range of acceptable accounting policies, some corporations choose to account aggressively and others conservatively. To repeat, management has quite a bit of latitude, largely because accounting rules cannot be written—and probably should not be written—so tightly as to eliminate this discretion.

Aggressive policies typically lead to earlier recognition of higher profits. However, these policies can lead to periodic "surprises" resulting in restatements and significant write-offs; such "surprises" almost always have a negative impact on profit. Recall from Chapter 3 the requirement that accountants lean toward *conservatism* in resolving valuation question. The rub is that conservatism is largely in the eye of the beholder; what is a conservative valuation to person A may strike person B as a quite aggressive valuation. Where to operate along the policy spectrum from very aggressive to very conservative is a matter of choice.

The culture of the corporation and the ethical framework of its management and board of directors typically drive that choice. At the risk of "preaching," let me outline a number of widely accepted truisms about corporate ethics and culture:

1. The tone is set at the top. A CEO who is inclined to look the other way when aggressive valuations are pursued or misbehavior occurs quickly causes that attitude to spread throughout the organization. And his or her "talking the talk" of ethical behavior will be ignored. The CEO who "walks the walk," insisting on conservative valuations and coming down hard on team members who break the rules, reinforces the corporation's code of ethics.

2. Single-minded obsession with "making the numbers"—that is, achieving announced revenue and profit targets—may tempt managers to stretch too far in accelerating revenues and delaying expenses.

3. Some CEOs do not want to hear bad news, particularly bad news that affects profit or financial position; a few may be inclined "to kill the messenger" who brings such news. The wise CEO demands that he or she receive bad news without delay and that any unfavorable financial impact be recognized immediately.

4. Incentive compensation plans that tie management bonus payments to short-term financial results can create almost irresistible temptations to accelerate revenues and/or delay expenses. Boards of directors are wise to gear bonus plans to longer term performance; doing so tends to align the interests of executives with those of shareholders.

5. A deeply imbedded culture of compliance is the best insurance for strong corporate integrity and sound financial reporting. Such a corporate culture insists on accurate and fair reporting, strict avoidance of conflicts of interest, genuine concern for the best interests of customers and shareholders, and compliance with the spirit as well as the letter of both corporate policies and the law.

Even in the best-managed companies, however, occasional ethical lapses on the part of one or a small group of employees may result in fraudulent activities. All corporations need to have in place control processes and procedures to guard against and to detect embezzlement, bribery, falsification of documents, and misrepresentations to customers, vendors, taxing and regulatory agents, and others.

ACCOUNTING RULES

In addition to the accounting guidelines discussed in Chapters 3 and 4, a host of reasonably specific accounting rules must be followed. The purpose of these rules is to insure some (but not perfect) uniformity across companies' financial reports. This uniformity improves comparability between financial statements of Company A and those of other companies in the same industry.

A small sample of such accounting rules is as follows:

- Accounting for currency fluctuations (at multinational companies).
- Amortizing capitalized software.
- Amortizing goodwill.
- Accounting for so-called finance lease (as opposed to rental) arrangements.
- Calculating earnings per share.
- Accruing for employee retirement benefits, including pensions, medical benefits, and other deferred income.
- Accounting both for newly issued employee stock options and for outstanding and unexercised stock options.
- Accounting for deferred income taxes.

Who promulgates these rules? The primary rulemaking body is the **Financial Accounting Standards Board (FASB)**, a nongovernmental board established in 1973 and located in Connecticut. While the FASB has a broad mandate, it is dedicated to an open and participative rulemaking process. Before a new rule is issued, a draft of the rule is widely circulated among corporations and accountants for comments and suggestions. Despite this participatory process, some FASB rules have been highly controversial. Incidentally, other countries with advanced economies have similar rulemaking bodies and attempts are made to minimize rule differences among all of them.

The Securities and Exchange Commission (SEC), an agency of the U.S. federal government first mentioned in Chapter 5, also has the authority (and capacity) to set accounting rules for those companies whose securities are traded publicly. The SEC has chosen to accept the authority of the FASB in almost all instances of general rules. Nevertheless, the SEC has the authority to challenge and require changes in accounting policies and financial statement presentations of so-called "registrants", that is, companies whose securities trade in public markets (or who seek to have their securities so traded) and who are thus required to file financial statements on a regular basis with the SEC.

The full set of rules governing how companies account and how they present their financial information to the public is widely referred to as **GAAP** (pronounced "gap"): **generally accepted accounting principles**. These policies, of course, evolve over time. If a company chooses to ignore a GAAP policy in determining its net income, the company must also show what its net income would be under GAAP. Occasionally, a company feels strongly that GAAP earnings are in some way misleading to investors and elects to report both GAAP and non-GAAP earnings and EPS.

AUDITS AND AUDITORS

Most companies—and not-for-profits as well—have their fiscal-year financial statements audited; corporations registered with the SEC are required to do so. Independent certified public accounting firms conduct audits. Four major public accounting firms—down from eight a few years ago because of consolidations and one failure—do business throughout the entire world (using affiliated firms in certain countries). These four firms audit a very high percentage of the largest U.S. and global companies. Each of them has revenues measured in billions of dollars! But they do not have all the auditing business. A second tier of large, well-respected firms is growing in importance as the number of elephant-size firms has shrunk. Finally, innumerable small shops perform audits and quasiaudits for less prominent corporations, partnerships, and not-for-profits.

Auditors are not responsible—and to remain independent must not be responsible—for doing the accounting for clients or constructing their income statements. Corporate management carries those responsibilities. Auditors conduct various tests and investigations to independently verify the values of certain assets and liabilities (for example, accounts receivable, accounts payable, inventory, fixed assets, loans payable, and accrued liabilities). They also review their clients' accounting systems and procedures to assure themselves that these are capable of recording both accurately and in a timely manner the many transactions that occurred during the year. They review all adjusting entries and assure themselves that the clients' accounting policies are consistent with GAAP.

At the end of this process, the auditors "sign off" in a report (often called the auditor's **opinion**) typically addressed to the client's board of directors. Over the years these statements have become quite stylized; stock phrases appear in statement after statement. The following stock sentences, drawn from a couple of recent annual reports, reveal some important points:

- "These financial statements are the responsibility of the company's management; our responsibility is to express an opinion on these financial statements based on our audit."
- "In our opinion, the financial statements present fairly, in all material respects, the financial position and the results of operations and cash flows for the years ended __, in conformity with U.S. generally accepted accounting principles."

Note in the second sentence the references to GAAP and to materiality. The opening phrase of the second sentence is the key: "in our opinion." That phrase puts the auditors' professional competence on the line; should it later come to light that their audit tests and procedures were inadequate and therefore not sufficient to support this "opinion," the auditing firm may be sued. The likelihood of a suit is substantially increased if the client's stock price takes a nosedive when, sometime later, overstated profits are revealed and the auditors did not catch that overstatement.

These sentences were taken from so-called "clean opinions." In some instances the auditors take exception to some policy or condition at the client company. Obviously, a reader of financial statements should pay very close attention to any "exception" statement within the opinion.

Auditors do not assure the directors and shareholders that no fraud occurred at the company. Auditors are typically quite careful to remind clients that, while they will report fraud if they find it, they cannot guarantee that they will uncover it. When several members of management conspire and collude in fraudulent actions, that fraud can be exceedingly difficult to detect, particularly in the short term.

Auditors now are also called upon to render opinions on the adequacy of the company's accounting controls. This new requirement has substantially increased the work required of auditing firms. Of course, this additional work is reflected in very much higher fees charged to clients.

Auditors are required to be "independent" of their clients. Therefore, for example, they cannot be hired by the client to operate its accounting system. Following the financial scandals in the very early years of this century, it was decided that audit firms compromise their independence when they earn substantial consulting fees from audit clients. As a result, the major auditing firms have spun off their very large consulting practices into separate business entities.

Auditors are in an interesting and troubling conflict of interest position. While the board of directors, rather than management, officially appoints the audit firm, in practice directors typically accept management's recommendation as to which firm to appoint. Audit fees are high; audit firms hate to lose clients, particularly large ones. Now suppose that management and the auditors have a disagreement as to a certain accounting policy or just how high a reserve should be or when it should be recognized. If management subtly threatens to change audit firms unless management gets its way, will the audit firm's resolve tend to weaken? To level the playing field somewhat, the SEC has established rules that now require that any auditor–management disagreement be thoroughly disclosed, a disclosure that most managements want to avoid.

The financial scandals mentioned earlier revealed some shoddy work by some audit firms. As a result, the SEC created an organization to establish standards for auditing and to audit the auditors: the **Public Company Accounting Oversight Board (PCAOB)**. It is still too early to assess how effective this new organization will be.

REGULATORS

A thorough discussion of the regulators and the regulations pertaining to publicly traded companies is beyond the scope of this book, but a couple of comments are appropriate.

The SEC (Securities and Exchange Commission) was established in the 1930s to curb abuses in the financial markets. Note its name: securities *and* exchange(s). The SEC regulates both the securities issued by corporations and the exchanges where those securities are traded among shareholders. The SEC's constituency is the investing public. Its work is the key to maintaining shareholder confidence in the fairness of the markets and the securities traded therein.

The SEC insists on **full** and fair **disclosure** of information relevant to making sound investment decisions. (See the discussion of disclosure in Chapter 5.) It is particularly keen that the risks that the company is exposed to be fully disclosed and discussed. Of course, the SEC cannot insist that investors actually understand and utilize the disclosed information. Investors are quite free to make stupid decisions. The SEC offers no opinions on whether a particular investment is "good" or "bad" or whether the price of a security is too high or too low.

When a private company seeks to "go public"—for the first time sell its securities to the public—the company must file with the SEC a very comprehensive document, called an S-1, that reveals great detail about the company, its management, the risks it faces, and its financial statements for the past number of years. The SEC examiners ask themselves: has the company revealed all information relevant to an investor making an informed decision whether to purchase the security and is all disclosed information accurate and not misleading? That is a tough, dual requirement for a company that aspires to go public. Unsurprisingly, S-1 documents have over the years become longer and longer, and progressively more complicated, as companies guard against being sued for failure to make full disclosure. Unfortunately, the longer the disclosure documents become, and the more they are written in legalese rather than the King's English, the less likely are investors to wade through the detailed and convoluted disclosure.

Once public, with its securities trading in the market, the company then must make periodic reports—including quarterly and annual financial reports—to the SEC. In recent years, more and more companies have elected to bind their annual filings (called **10-K**s) into their annual shareholder reports. To underscore the problem of excessive disclosure, let me note that the 10-K report for a large and prominent insurance company is 243 pages long. Ask yourself who other than the company's lawyers and chief financial officer has actually read each of those pages—probably not many shareholders.

Many industries are regulated by other state and federal government agencies: public utilities, broadcasters, defense contractors, and so on. In some cases, the regulating agencies have insisted on certain accounting policies and presentations.

Finally, securities exchanges themselves promulgate certain rules—some of them related to financial information—that must be followed by companies whose securities trade on those exchanges. Thus, the New York Stock Exchange, the NASDAQ exchange, the London Stock Exchange, and others have a quasiregulatory function.

COOKING THE BOOKS

Readers of financial statements need to maintain a healthy skepticism—but, please, not undue cynicism—about the "truth" revealed in those statements. Financial statements are just an educated and faithful estimate of financial position at a moment in time and performance for a period of time. Even when conservative accounting policies are pursued and operational and financial controls are sound, subsequent events may have major impacts on the values of assets and liabilities:

- A pharmaceutical company's major drug is discovered to have major detrimental side effects; suddenly the company's liabilities soar.
- A consumer electronics company finds that its "hot" gaming device is rendered obsolete overnight by a competitor; it now must expect major product returns from its retailers—accounts receivable are overstated—and a good portion of the inventory is useless.
- An information technology company that develops large software systems for the financial industry discovers that a contract that management thought was 90% complete is in reality only about 60% complete; the loss that is now anticipated on this contract must now be included in the financial statements.

Those three examples do not involve cooking the books. They illustrate simply the "surprises" that crop up in business that can have major financial impacts. Cooking the books involves taking deliberate steps—not necessarily either fraudulent or illegal, but very frequently unwise—to "dress up" the balance sheet (enhance the apparent financial position of the company) or improve profits. No bright line exists, of course, to tell us when a company has wandered into "book cooking" territory!

We can categorize "book cooking" action as follows:

- Overstating assets.
- Understating liabilities.
- Overstating or accelerating revenues.
- Understating or delaying the recognition of expenses.

Remember that the flip side of overstating assets is often understating expenses or overstating liabilities. And the flip side of accelerating revenues is generally the overstatement of assets or the understatement of liabilities.

Let us consider a few examples in each category. I will avoid using company names to guard against libel suits!

Overstating Assets

A major communications company capitalized an expenditure that should have been expensed. Maintenance expenditures on its facilities were treated as if they were investments in new or upgraded facilities. This error increased the value of fixed assets and, importantly, avoided recognizing very large expenses, thus overstating net income.

Whenever a company fails to provide adequate reserves for adjusting the values of its assets, it is overstating assets and understating expenses. Japanese banks for many years refused to provide adequate reserves for loan losses (recall that loans are the primary asset of banks)—with the government's bank regulators' explicit or tacit approval. In many cases, the borrowers were not paying interest and the banks had little chance of obtaining principal repayment. The Japanese government recognized that fully reserving against these probable loan losses would render many banks insolvent (that is, push them into a position of negative owners' equity) and the government and the banks were unwilling to face up to such a financial crisis.

Certain long-term assets, particularly receivables, are best valued by the time-adjusted value method. Using an unrealistically low discount factor results in an overstatement of the asset.

If assets such as accounts receivable or cash are encumbered (that is, pledged in support of borrowing), this encumbrance should be clearly disclosed in the financial statements or in the accompanying notes.

Other examples include the following:

- Depreciating fixed assets over too long an estimated life.
- Providing inadequate allowances for doubtful accounts.
- Avoiding full recognition of impairment in goodwill values.
- Not reserving for inventory obsolescence that often accompanies the introduction of a new product model utilizing relatively few of the components of the displaced model.
- Using the gross method of accounting for sales (and accounts receivable) while offering customers compellingly attractive discounts for prompt payment.

Understating Liabilities

The most prominent example these days is the failure of many large and mature companies to recognize the full extent of their obligations to pay pensions to and medical insurance premiums for present and future retirees. In addition, some companies enter into deferred compensation agreements with senior managers without recognizing these obligations on the balance sheet and expensing them as the managers accrue the benefits.

Recall from the previous section that the use of too low a discount rate will overstate long-term assets valued by the time-adjusted method. In a parallel manner, too high a discount rate used to value long-term liabilities will result in an understatement of the liabilities.

Other examples include the following:

- Inadequate reserves (liabilities) to effect warranty repairs in future years.
- Failure to accrue vacation wages liabilities for salaried employees.

Overstate or Accelerate Revenues

A practice known as "stuffing the channels of distribution" involves requiring dealers and distributors to acquire unreasonable amounts of inventory toward the close of the manufacturer's fiscal year so as to accelerate into the current year sales that would more appropriately be made and recorded in the next year.

Companies who receive upfront fees from franchisees (e.g., fast-food restaurants) have had a tendency to include in current revenue too large a portion of those fees in the accounting period when the franchise contract is signed, without giving adequate recognition to the continuing servicing obligation that the franchisor has to the franchisees.

Companies selling to themselves—that is, recording as bona fide sales the shipment of goods to affiliated entities—have caused some financial scandals. Legitimate revenue is earned only in transactions with independent, arm's-length customers.

Understate or Defer Expenses

A practice that for years caused great concern among some accountants and financial analysts was the failure of most companies to recognize any expense in connection with granting stock options to employees. Obviously stock options have a value; if they did not, employees would be indifferent about receiving them. Only recently has an FASB pronouncement required that an expense be recorded at the time the option is issued. A remaining key challenge is agreeing on just how to value an option at the time of grant; it is that value that should be recognized as additional compensation expense at the time of grant.

SUMMARY

At the conclusion of this book, it is worth reminding ourselves again that financial statements are, at best, informed, reasoned, thoughtful, conservative estimates—but only estimates—of an organization's financial performance for a period of its life and its financial position as of a single moment of its life. Operations continue in the next accounting period and the only way that we could be certain of the organization's lifetime financial performance would be to cease operations, sell all the assets, pay off all the liabilities and see what is left!

The use of estimated values is essential; we can be appropriately skeptical about these values, but not cynical so long as we have confidence in the integrity and honest intentions of the organization's management.

I know a "money manager" who avoided buying Enron common shares throughout the years of their spectacular market appreciation. When asked how she had avoided the pressure to get on the bandwagon and invest in this "hot" stock, she replied that, based on her face-to-face interviews with Enron top executives, she concluded that she did not trust them. That insight trumped all the elegant financial analyses she had done. Just imagine, when the collapse of Enron proved her correct, what a hero she was to her clients!

NEW TERMS

Financial Accounting Standards Board (FASB)

Full disclosure

Generally accepted accounting principles (GAAP)

Opinion (of auditors)

Public Company Accounting Oversight Board (PCAOB)

10-K

Appendix: Scorekeeping at Not-for-Profits

Rest assured that **not-for-profit institutions**—colleges, universities, social welfare agencies, foundations, some hospitals, religious institutions, many research institutes—do not require an entirely different system of accounting or an unfamiliar set of financial statements. Double-entry bookkeeping, ledgers, charts of account, debits and credits are alive and well at not-for-profits. Their statements are close cousins to those of for-profit enterprises that we have been reviewing, but there are some differences.

HOW ARE NOT-FOR-PROFITS DIFFERENT?

Predictably, the statement differences are driven by the differences in institutional objective: profit for its owners versus service for its constituency. Although so-called not-for-profits do not seek to earn a profit, financial objectives play key roles in their operations. A frequent financial objective is at least to break even; that is, assure that revenues match or exceed expenses. But at other times and in other situations a not-for-profit may have the objective of building up its reserves to reduce its financial risk, or drawing down its reserves by a tolerable amount so as to expand its public services. Some not-for-profits (but very few profit-seeking companies) plan an orderly phase-out of operations once their objectives have been realized.

Not-for-profits, although typically incorporated, do not have shareholders. Think of their reserves—total assets less total liabilities—as being held in trust for the benefit of the public, typically that segment of the public—students, less fortunate citizens, parishioners, the general public—identified as their constituencies. A not-for-profit typically has a **board of trustees** (sometimes called board of directors) that has the fiduciary responsibility to see that the public trust is honored and fulfilled. This fiduciary obligation typically extends not only to serving its current clients (recipients of service) but also to servicing future clients; this obligation is often referred to as responsibility for intergenerational equity. The board also has an obligation to steward funds entrusted to the not-for-profit by its donors, who, of course, may be members of the constituency itself: art and music patrons or parishioners. Just as with for-profit corporations, not-for-profits can be wound up (dissolved), merged, or (very infrequently) sold. The board of trustees' responsibilities are in fact quite parallel to those of a for-profit board of directors: approve policy and direction for the organization and hire, supervise, and, if necessary, replace the management leader, who may carry the title of executive director, president, chancellor, head, chief executive, or some other.

The fundamental accounting equation is also alive and well at not-for-profits, but now of course the difference between total assets and total liabilities must carry a label other than shareholders' or owners' equity. There are no owners or shareholders. Instead the term "net assets" and sometimes the word "reserves" is used. Both labels are unfortunate, as they suggest that the difference between assets and liabilities is "spendable"; of course, it is not, any more than shareholders' equity is spendable. We will stick with the phrase "net assets" although I would prefer a label such as "retained and invested resources." The point is that so-called "net assets" are quite analogous to retained earnings in a for-profit environment.

Of course, the not-for-profit has no equivalent to "invested capital" at for-profit companies, as it has no investors to whom it sells shares of common stock. Unsurprisingly, statements of performance are not called Income Statements or Profit and Loss Statements, since profit and loss are irrelevant concepts in these organizations. It is not irrelevant, however, that in some accounting periods revenues exceed expenses and that in other periods the reverse condition obtains; those conditions are generally labeled in a very straightforward manner: "excess of revenue over expenses" and "excess of expenses over revenues," respectively. The income statement equivalent is typically labeled "Statement of Activities."

The balance sheet equivalent typically carries the happy label "Statement of Financial Position," exactly the phrase we used in Chapter 1 to define balance sheet! The cash flow statement is most frequently called exactly that!

GIFTS

As you know, in the U.S. gifts to not-for-profits are deductible from the donors' personal incomes when calculating their income taxes; this deductibility is subject to complex rules and limits and requires that the not-for-profit be qualified as a so-called 501(c)-3 corporation.

Gift support is all important to not-for-profits. But not all gifts are created equal. Two types of restriction are frequently imposed by donors: restriction as to *purpose* for which the gift may be used and restrictions as to *when* the gift may be spent.

Some gifts arrive without donor restrictions as to purpose and thus may be used at the not-for-profit's discretion. Other gifts are "designated" by the donor; they must be spent only for the designated purpose(s).

Further, some gifts are immediately expendable and others are restricted as to when they can be spent. These latter gifts are placed into the institution's **endowment**. An all-important feature of many—but certainly not all—not-for-profits is the buildup, management, and use of this endowment. A slight paraphrase of the dictionary definition of endowment is "a gift or bequest [gift by will at death] that provides future income for an institution."

Once in endowment, various restrictions may apply. Some endowment funds may never be spent down to zero (except under desperate circumstances). Others may be spent only after certain milestones or conditions are met or before or after specified future dates. A major factor complicating the financial reporting of not-for-profits is the need to track these restrictions.

ENDOWMENT

Endowments are augmented both by (a) new gifts designated by their donors as endowment gifts, and (b) reinvestment of earnings on the endowment. The prudent not-for-profit invests endowment funds in a variety of investment vehicles, including common stocks (both domestic and international), bonds, real estate, cash and cash equivalents and, in some cases, so-called **alternative investments** such as venture capital and hedge funds. The U.S. income tax laws provide that earnings on not-for-profits' endowments are not taxed.

So-called "true" **permanently restricted endowment** consists of those funds designated by their donors as intended to benefit the organization "in perpetuity"—forever! The organization's ability to spend such "true" endowment funds is severely limited. **Temporarily restricted endowment** consists of gifts on which the donor has placed restrictions prohibiting their current use; as these temporary restrictions are fulfilled or lapse (by the passage of time or achievement of milestones), the funds become available to be spent. Quasiendowment (sometimes called "funds functioning as endowment") consists of monies so designated by the institution's board of trustees. **Unrestricted endowment** consists of gifts so designated plus earnings on "true" endowment. So-called **term endowment** consists of funds designated by the donor to be spent down to zero over a specified number of years seldom exceeding 10. If all this sounds complicated, it is!

The institution's board of trustees has the fiduciary responsibility for investing these endowment funds in a prudent manner, balancing risk of loss with potential for gain. It also must balance service to current clients with its obligation to intergenerational equity (service to future clients). Typically, the board adopts an **endowment spending rate** policy; frequently this target rate is equal to the anticipated total return on endowment (dividends, interest, and market appreciation) less the amount that must be reinvested in order to maintain the endowment's purchasing power (i.e., less a relevant inflation rate). This target is designed to maintain the endowment's purchasing power over the long term, although not necessarily in any particular year. For example, if the board anticipates that annual earnings from the chosen mix of investments will average 9% and the average relevant inflation rate (not necessarily the consumer price index or gross domestic product deflator) will be 4% per year, the institution can withdraw and spend 5% of the endowment value annually to support its current operations. Because the values of endowments fluctuate year to year with volatility in investment returns, most institutions use a "smoothing" formula to moderate the volatility of the annual withdrawals.

As a matter of fact, endowed institutions in the United States have, on average, been quite conservative in setting their endowment spending rates; this conservatism has resulted in increases in total endowment funds well in excess of the sum of inflation plus new endowment gifts. For example, at Harvard University extraordinarily handsome investment returns coupled with modest spending rates have resulted in its endowment ballooning to $26 billion by early 2006.

REVENUES AND EXPENSES

Expenses at not-for-profits are not materially different than those at for-profits: salaries and wages, employee fringe benefits, rent, utilities, depreciation, insurance, and other operating expenses. Matching of expenses to accounting periods is just as important to accurate accounting for not-for-profit operations as it is for for-profit operations.

A not-for-profit typically receives (earns) a mix of revenues of three kinds: (a) fees paid by those to whom it provides services (e.g., tuition from students, attendance charges, below-market-rate charges for food or room, ticket sales); (b) gifts, often carrying various restrictions imposed by donors; (c) payout (draw) from the institution's endowment. The presence and relative importance of each source vary widely across the broad community of not-for-profits.

As hinted above, accounting for gifts is complicated. Frankly, it is also somewhat inconsistent among not-for-profits. In general, gift revenues consist of the following:

- Gifts received this year that are immediately available to support current operations; these are generally referred to as currently expendable gifts. Alumni funds at colleges and universities and annual campaigns at religious institutions, hospitals, and social welfare agencies are examples.

- The portion of restricted gifts received in prior years on which the restrictions have now lapsed or been fulfilled and are thus available to be expended.

- Gifts received with donor-imposed restrictions that preclude their immediate use.

- Gifts restricted to the endowment.

The first two categories represent current revenue; the last two obviously must be accounted for but not as current revenue.

Furthermore, the institution's endowment may have earned **gains** or **losses** from investment activities, and those gains or losses may have been "**realized**" or remain "**unrealized**." Gains or losses are "realized" only when the security or other investment is sold; when securities are revalued—marked to market—any gains or losses exist only on paper and thus are considered "unrealized." These gains and losses, both realized and unrealized, also need to be recognized in the financial statements.

In a moment we will look at some examples that bring some clarity to this accounting puzzle.

ASSETS AND LIABILITIES

Liabilities at not-for-profits are not unlike those at for-profits: accounts payable, salaries and benefits payable, loans payable (short and long term), accrued liabilities, and deferred revenues of various kinds. Most not-for-profits are able to borrow some amount from banks on a short-term basis and many are able to gain mortgages collateralized by their real estate holdings.

Well-established not-for-profits, particularly those with substantial endowments, are able to issue bonds in the public market. Educational institutions, hospitals, and some other not-for-profits are able to issue (technically through a government agency) tax-free bonds, that is, bonds on which the interest payments are not taxable to the bondholders (similar to bonds issued by municipalities and public school districts). This tax-free feature permits the institution to borrow at lower interest rates. Once again, the federal government confers this tax advantage to only certain categories of not-for-profit institutions (e.g., not religious institutions) and it does so because the institutions' activities are presumed to be in the public interest.

The asset side of the Statement of Financial Position differs from those of for-profit institutions primarily because of the inclusion of endowment assets, often the dominant asset owned. Otherwise the not-for-profit owns cash and cash equivalents, accounts receivable (for fees earned but not yet received), supplies inventories, prepaid assets, and fixed assets. Goodwill and other intangibles seldom amount to much.

How should the not-for-profit account for gift pledges: donors' promises to make future gifts? Until about a decade ago, gift pledges were ignored until the pledge was in fact fulfilled (i.e., paid by the donor). Now the prevailing accounting rules require that if a pledge is firm and documented—something considerably more solid than a donor's casual statement of intent—it must be counted as revenue in the period when the firm pledge is received. Some pledges provide for payments over a number of years; thus not all unpaid pledges should appear as *current* receivables. As with accounts receivable, not-for-profits often—and wisely—establish an allowance (in a contra account) for pledges that will ultimately prove to be uncollectible, particularly, since not-for-profits are often reluctant to take legal action to force the payment of pledges.

EXAMPLE: AVENIDAS

Avenidas is a California social welfare agency serving the social and health needs of older adults in a set of communities on the San Francisco peninsula. Its financial statements for the years ended June 30, 2005 and 2004 are shown in Exhibits A-1, A-2, and A-3: the Statement of

Exhibit A-1: *Avenidas Statement of Financial Position ($ thousands)*

	JUNE 30	
	2005	2004
Assets		
Cash and cash equivalents	$2,425	$940
Investments in marketable securities	16,891	15,348
Accounts receivable, net	201	259
Pledges receivable	981	3,389
Deposits and prepaid expenses	50	52
Property and equipment, net	2,769	792
Total assets	$23,317	$20,780
Accounts Payable and Accrued Liabilities	$585	$189
Net Assets	22,732	20,591
Total liabilities and net assets	$23,317	$20,780

Financial Position (equivalent to a balance sheet), the Statement of Activities (analogous to an income statement), and the Statement of Cash Flows.

Avenidas' Statement of Financial Position is obviously dominated by a very handsome amount invested in marketable securities. Its liabilities are minimal and thus its Net Assets are substantial and growing. Note that Net Assets have increased by approximately the amount of the surplus shown on the Statement of Activities (Exhibit A-2)—just as net income adds to retained earnings in a for-profit company.

Avenidas' published financial statements also classify the Net Assets at June 30, 2005 as follows (in $ thousands):

Unrestricted	$14,563
Temporarily restricted	8,382
Permanently restricted	372
Total Net Assets	$23,317

With almost two-thirds of its net assets unrestricted, and virtually all the remainder only temporarily restricted, Avenidas has great flexibility as to the use of its funds. Also, less than 12% of its assets are tied up in property and equipment.

Exhibit A-2: *Avenidas Statement of Activities ($ thousands)*

	YEARS ENDED JUNE 30	
	2005	2004
Public Support and Operating Revenues		
Program fees	$1,036	$954
Community contributions	666	731
In-kind professional services	236	206
Government support (city, county, federal)	510	564
Total	$2,448	$2,455
Expenses		
Program services	2,705	2,560
Management and general	329	296
Fundraising	351	368
Total	3,386	3,224
Surplus (deficit) before investment items	(938)	(769)
Investment income	521	335
Realized and unrealized gain on investments	941	1,630
Restricted donations to endowment	1,616	3,814
Avenidas surplus (deficit)	$2,140	$5,010

What about still other balance sheet ratios, those we use to analyze for-profit institutions? Avenidas is extremely liquid: its cash balance is over four times its total liabilities. Its debt leverage is zero.

The Statement of Activities is a bit more difficult to analyze. A casual reader might incorrectly conclude that Avenidas has had two difficult years, recording substantial deficits (before investment items) equal to 38 and 31% of current revenue. Or, one might conclude that Avenidas had two great years, since the "bottom line" shows a very large surplus in both years due to impressive totals of new gifts to endowment in each year. Both interpretations would, unfortunately, be misleading.

Given Avenidas' substantial endowment, it has no need to—indeed it should not—restrict its expenditures to the sum of current gifts plus program and in-kind service fees; that is, it should expect to generate a "deficit" at the line labeled "Surplus (deficit) before investment items." But the "bottom line" is equally misleading. Gifts to endowment may not be spent when received, while realized and unrealized gains on endowment will fluctuate considerably

Exhibit A-3: *Avenidas Statement of Cash Flows ($ thousands)*

	YEARS ENDED JUNE 30	
	2005	2004
Cash Flows from Operating Activities		
Change in net assets	$2,140	$5,010
Adjustments to reconcile to net cash		
Depreciation & amortization	111	117
Realized/unrealized gain on investments	(941)	(1,630)
Decrease in allowance for doubtful accounts	(1)	(3)
Changes in operating assets/liabilities		
Accounts & pledges receivable	2,467	(3,208)
Other	399	73
Net cash provided by operating activities	4,175	359
Cash Flows from Investing Activities		
Purchase of marketable securities	(4,813)	(12,973)
Proceeds from sale of marketable securities	4,212	12,962
Purchase of property and equipment	(2,089)	(192)
Net cash used in investing activities	(2,690)	(202)
Net increase in cash and cash equivalents	$1,485	$157

from year to year. Thus, neither the "deficit" figure nor the bottom-line "surplus" figure gives us an accurate read on whether operations over these years were financially satisfactory and sustainable.

The whole purpose of building an endowment at Avenidas is to provide funding for program services. Here, the average of the two year-end "Investments in marketable securities" values is about $16.1 million, and the $938,000 deficit is just over 5.8% of that average. Prudently invested, this endowment should be able to earn over the long term average annual total returns of, say, 10%, but that is unlikely to be quite enough to cover both (a) nearly 6% payout for operations and (b) inflation. Accordingly, Avenidas is continuing aggressive fundraising to augment its endowment.

As noted in Chapter 5, total return on securities investments is equal to the sum of the current return (interest and dividends) and market price appreciation. In 2005 Avenidas's current return was $521,000, about 2.4% of the average value of endowment. If this seems low, remember that yields on common stocks in 2005 averaged well below 2%. On the other hand, market price appreciation, both realized and unrealized, was about 4.3% of average endowment value. Total return on the endowment, then, was about (4.3 + 2.4) 6.7% in 2005—not a great year for market returns, and a return not sufficient to cover both the inflation rate plus Avenidas' operating deficit. But in the previous year, Avenidas' total return on endowment was well above the target long-term rate.

While the financial statements themselves do not reveal Avenidas' spending policy, as established by its board, Avenidas' executive director reports that the policy is "to contribute annually to operations an amount equal to the amount transferred the previous year inflated by the Consumer Price Index, provided that the resulting amount is between four and five percent of a rolling 12-quarter average asset value of the endowment." This policy appears to be sensible, steady, and conservative.

All in all, both the financial position and the activities of Avenidas appear strong over this 2-year period.

Incidentally, accounting rules wisely require not-for-profits to itemize the total amount spent on fundraising. This disclosure has reduced the number of questionable not-for-profits who spend on fundraising a very large percentage of the funds raised. Thus, a pertinent ratio for not-for-profits is fundraising expenses to total funds raised. Avenidas' ratio averaged about 10.5% over these 2 years, an excellent ratio.

The Avenidas' Statement of Cash Flows does not add much to our knowledge of Avenidas, but it does provide a bit of additional practice on the art and science of reading this third important statement.

Once again, the news here is not in the bottom line, although this year's increase in cash is impressive. Note that the typical third section of cash flow statements—cash flows from financing activities—does not appear because Avenidas has done no financing. We see that Avenidas has made (and may continue to make?) large investments in property and equipment; indeed Avenidas is building a new senior center. Gift pledges grew by a large amount in 2004 due to a major fundraising campaign, and almost $2.5 million of that pledge balance was received (added to cash) in 2005.

Why is "realized/unrealized gain on investments" shown in parentheses? Because these gains appeared as revenue in the Statement of Activities but added to endowment value and did not result in increased operating cash; this is just the reverse of the situation with depreciation and amortization, which appear as expenses but do not consume cash.

The cash flow from investing activities section contains some very large dollar amounts, which to my mind serve more to confuse than enlighten. The net of these big numbers is that Avenidas invested a net $601,000 in marketable securities in 2005 and $11,000 in 2004. The rest is just investment churn!

Exhibit A-4: *Stanford University Statement of Financial Position, August 31, 2005 ($ millions)*

Assets	
Cash and cash equivalents	$434
Accounts receivable (including students), net	316
Pledges receivable, net	473
Inventories	54
Loans receivable	269
Investments	15,132
Plant facilities, net	2,354
Collections of works of art	–
Total assets	$19,032
Liabilities	
Accounts payable and accrued expenses	$501
Notes and bonds payable	1,266
Other liabilities	1,534
Total liabilities	3,301
Net Assets	
Unrestricted	11,547
Temporarily restricted	560
Permanently restricted	3,623
Total net assets	15,730
Total liabilities and net assets	$19,032

ANOTHER EXAMPLE: STANFORD UNIVERSITY

Stanford University's fiscal year ends on August 31, just before the start of its new academic year. Its financial statements are, unsurprisingly, quite complex and made more so by the fact that the university owns and operates (as a subsidiary) a hospital that accounts for almost 40% of consolidated revenues and expenses. Exhibits A-4 and A-5 show abbreviated financial statements for just the university (not consolidated with the hospital). To my mind, these statements are somewhat more useful than those of Avenidas.

As with Avenidas, Stanford's Statement of Financial Position is dominated by the value of its endowment. Note that the aggregate value of donor pledges subject to definitive agreements is listed as an asset, reduced by an unspecified allowance for uncollectible pledges. The footnotes

Exhibit A-5: *Stanford University Statement of Activities, year ended August 31, 2005 ($ millions)*

Revenues	
Student income (net of $137 of financial aid)	$356
Sponsored research support	973
Current year gifts in support of operations	144
Net assets released from restrictions	82
Investment income distributed for operations	514
Program fees and other income	558
Total revenues	2,629
Expenses	
Salaries and benefits	1,469
Depreciation	192
Other operating expenses	839
Total expenses	2,499
Excess of revenues over expenses	$130

indicate that the value of these pledges has been discounted for both time (the great bulk of these are due 1 to 5 years in the future) and possible uncollectibility; the discount is substantial, 25%. Stanford's art collections, all received or financed by gifts, are not valued here, presumably because Stanford has no intention of selling them; the label on this statement, however, reminds the reader that this is an important asset class for the university. The footnotes also tell us that the aggregate original cost of the campus plant facilities was $4.2 billion and thus these facilities are about 40% depreciated.

Stanford's borrowing in the public bond markets totaled nearly $1.3 billion at the end of the fiscal year. As a percentage of plant facilities, investments, or net assets, this borrowing is small. The university holds cash nearly equal to its current liabilities and accruals. We can safely conclude that the university is very liquid and is financed in a conservative manner.

Interestingly, in neither the Avenidas nor the Stanford Statements of Financial Position are subtotals provided for current assets and current liabilities.

A major research university is obviously not a small enterprise, even ignoring its hospital operations: revenues approaching $3 billion annually, total assets of almost $20 billion, and net assets approaching $16 billion. Its donors have over the years been enormously generous to the university, beginning with Senator and Mrs. Leland Stanford!

The Statement of Activities (Exhibit A-5) goes on to record other changes in net assets (i.e., net worth, if this were a for-profit company), as follows:

Unrestricted net assets activity	
Investment gains not included in operations	$ 2,048
Other	109
Net change in unrestricted net assets	2,157
Temporarily restricted net assets activity	
Gifts and pledges, net	214
Net assets released to operations	(82)
Other	(46)
Net change in temporarily restricted net assets	86
Restricted net assets activity	
Gifts and pledges, net	243
Investment gains	133
Other	30
Net change in restricted net assets	406
Net change in total net assets	$ 2,650

Some of the relationships here may surprise you; others will not. Net tuition is only about 17% of total revenues and financial aid totals over 25% of gross student fees. Sponsored research support, primarily from the federal government, is nearly three times net student income; Stanford has large and productive science, engineering, and medical faculties who are successful in winning competitive research grants. Investment income distributed (from the endowment) in support of operations is nearly 1.5 times student net tuition and fees. Total net assets grew by over $2.6 billion or slightly over 20% in 2005. Thus, with a positive surplus in operations, wonderful success in attracting new gifts, and a strong record of investment gains, Stanford had a whale of a financial year in 2005!

We are not surprised that 60% of total expenses are comprised of salaries and benefits; teaching and research are labor-intensive activities. Classification of expenses by major activity—say, teaching, research, administration, and fundraising—might be more enlightening than the classification that Stanford provides, but this additional information is available in the footnotes. While Avenidas spells out on its Statement of Activities its expenditures on fundraising, we have to go to the footnotes to determine that Stanford spent on fundraising (what Stanford calls development) about $55 million, less than 10% of the more than $600 million it received during the year in new gifts and pledges.

I will spare you Stanford's cash flow statement because in fact it provides very little additional information.

In four important ways, Stanford's financial statements presentation differs from that of Avenidas. I find Stanford's presentation substantially more illuminating:

1. Stanford indicates, while Avenidas does not, the effect of its policy on endowment payout—see the line item "investment income distributed for operations." This amount includes the current dividend and interest income earned by the endowment plus enough of the accumulated realized and unrealized market value gains to total the desired payout, approximately 4.75% of the "smoothed" endowment value. Avenidas reports both current income and market appreciation but leaves to the reader to decide if these combined amounts were sufficient to cover the "deficit" in operations.

2. The detail on the changes in net assets that appears toward the end of the Statement of Activities (following the "excess of revenues over expenses" line) is useful.

3. Stanford's net assets are classified as to restrictions: unrestricted, temporarily restricted, and permanently restricted. Its Statement of Activities shows as revenue in 2005 the amount of temporarily restricted endowment on which restrictions have been lifted or have expired during the year 2005. Avenidas also supplies this information but in a far more complicated manner that is not reproduced here.

4. Stanford shows explicitly that investment gains on endowment (both restricted and unrestricted endowment) become themselves unrestricted. It is for this reason that net asset values are dominantly "unrestricted" (over 75% of total net assets) despite the fact that at the time gifts are made to endowment they are restricted (that is, "true" endowment).

UNRELATED BUSINESS INCOME

Before we complete the discussion of not-for-profit financial scores, I should mention that not infrequently not-for-profits engage in activities that do not qualify within their 501(c)-3 exemption. That is, they engage in activity that is so analogous to for-profit activities that any resulting "net" (revenue less expenses) is taxable. Such activities are not illegal but they are unrelated and thus any "net" is described as **Unrelated Business Income**; the not-for-profit must be very careful to separate the accounting for such activities and to remit appropriate taxes thereon.

ACCOUNTING PRINCIPLES FOR NOT-FOR-PROFITS

You surely see that the accounting principles that we discussed in the first four chapters apply equally well to not-for-profits:

- Conservatism.

- Consistency.

- Matching.

- Financial assets (such as endowments) are marked to market.

- Cost values predominate.

- Valuations must be feasible, reliable, timely and free from bias.

- Realization.

- Going-concern assumption.

- Materiality.

- Use of estimates.

Interestingly, about a decade ago the not-for-profit accounting rules changed in several ways, but one significant change has, to my mind, both reduced the usefulness of the resulting financial statements and moved away from conservative valuations. This change was to treat firm pledges of gifts as revenue (with, of course, the pledge shown as an account receivable). Remember that giving is a voluntary act, and pledges of gifts are, in practice, closer to "statements of intent" than "contracts." To my mind, then, current practice overstates both revenue and assets for growing not-for-profits. Pledges seem to me more analogous to customer purchase orders than they are to "sales/revenue." Thus, I should think that both the realization principle and the conservatism principle would argue that gifts should be recorded as revenue only when actually received in cash (or equivalent), and not when promised.

The major complication in not-for-profit accounting vis-à-vis for-profit accounting is the requirement that net assets be segregated into three buckets—unrestricted, temporarily restricted, and permanently restricted—and then any movements among these buckets (particularly the movement of temporarily restricted funds to unrestricted as restrictions are met or lapse) be traced.

SUMMARY

As you switch from analyzing for-profit financial statements to those of not-for-profit enterprises, bear in mind the following:

- Net assets at not-for-profits are generally analogous to net worth at for-profits; in both cases the value equals total assets minus total liabilities.

- Endowment arises from past gifts to the not-for-profit and comes in three flavors: unrestricted, temporarily restricted, and permanently restricted. For-profits have no analogous asset.

- Not-for-profits generally devote to current operations all current earnings on their endowment and an additional appropriation of past investment gains (realized or unrealized); the amount of that appropriation is typically a function of the payout policy established by the board of trustees. However, if that appropriation is insufficient to cover the operating deficit, the remaining deficit consumes additional unrestricted endowment assets.

- Returns—dividends, interest, realized gains, and unrealized gains—on permanently restricted and temporarily restricted gifts become themselves unrestricted.

NEW TERMS

Alternative investments

Board of trustees

Endowment

Endowment restrictions: permanent, temporary, unrestricted

Endowment spending rate

Net assets

Not-for-profit institutions

Realized (and unrealized) gain/loss

Permanently restricted endowment

Term endowment

Temporarily restricted endowment

Unrestricted endowment

Unrelated business income

Index

Author Biography

Mr. Henry E. Riggs is the Founding President Emeritus of Keck Graduate Institute of Applied Life Science, founded in 1997 in Claremont, CA. From 1988–97 he was President of Harvey Mudd College. From 1974–88 he was the Thomas Ford Professor of Engineering Management at Stanford University, where he was also Vice President for Development, 1983–88. Earlier in his career he held officer positions in two high technology companies. Mr. Riggs holds a B.S. degree in Engineering, an MBA (with high distinction) from Harvard University, and honorary doctorates from Harvey Mudd College and Keck Graduate Institute. He is the author of five books and numerous articles and papers.